Linear and Integer Programming Made Easy

T. C. Hu • Andrew B. Kahng

Linear and Integer Programming Made Easy

Springer

T. C. Hu
Department of Computer Science
and Engineering
University of California at San Diego
La Jolla, CA, USA

Andrew B. Kahng
Department of Computer Science
and Engineering
University of California at San Diego
La Jolla, CA, USA

ISBN 978-3-319-23999-6 ISBN 978-3-319-24001-5 (eBook)
DOI 10.1007/978-3-319-24001-5

Library of Congress Control Number: 2015950475

Printed on acid-free paper

This Springer imprint is published by Springer Nature
The registered company is Springer International Publishing AG Switzerland

Introductory Map of the Book

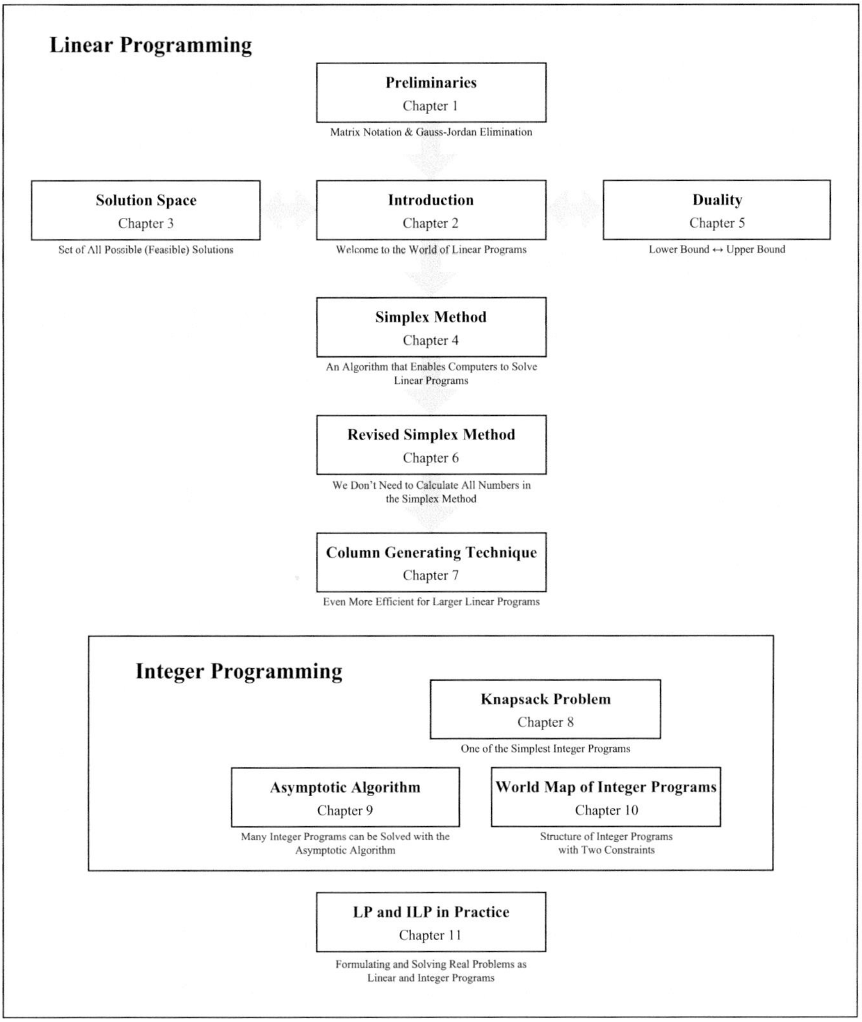
Linear Programming
Preliminaries
Chapter 1
Matrix Notation & Gauss-Jordan Elimination
Solution Space
Chapter 3
Set of All Possible (Feasible) Solutions
Introduction
Chapter 2
Welcome to the World of Linear Programs
Duality
Chapter 5
Lower Bound ↔ Upper Bound
Simplex Method
Chapter 4
An Algorithm that Enables Computers to Solve Linear Programs
Revised Simplex Method
Chapter 6
We Don't Need to Calculate All Numbers in the Simplex Method
Column Generating Technique
Chapter 7
Even More Efficient for Larger Linear Programs
Integer Programming
Knapsack Problem
Chapter 8
One of the Simplest Integer Programs
Asymptotic Algorithm
Chapter 9
Many Integer Programs can be Solved with the Asymptotic Algorithm
World Map of Integer Programs
Chapter 10
Structure of Integer Programs with Two Constraints
LP and ILP in Practice
Chapter 11
Formulating and Solving Real Problems as Linear and Integer Programs

Preface

The subject of linear programming was discovered by Dr. George B. Dantzig, and the subject of integer programming was discovered by Dr. Ralph E. Gomory. Thousands of papers and hundreds of books have been published. Is there still a need for this book?

The earlier algorithms for integer programming were based on cutting planes. In this book, we map non-basic columns into group elements to satisfy congruence relations. The first six chapters of the book are about linear programming. Then, Chapter 8 introduces the knapsack problem which is known to have time complexity of $O(nb)$, where n is the number of types of items and b is the capacity of the knapsack. We present a new algorithm which has time complexity $O(nw)$, where n is the number of types of items and w is the weight of the *best* item, that is, the item with the highest ratio of value to weight.

The unique contents of this book include:

1. The column generating technique for solving very large linear programs with too many columns to write down (Chapter 7)
2. A new knapsack algorithm with its time complexity $O(nw)$, where n is the number of types of items and w is the weight of the best item (Chapter 8)

The knapsack problem highlights two striking features of an integer program:

1. The optimum integer solution has a periodic structure.
2. The percentage of integer programs that cannot be solved by the group method becomes smaller and smaller as the right-hand side becomes larger and larger.

Thus, we explain the two features in detail and devote all of Chapter 9 to the asymptotic algorithm for integer programming; we present "The World Map on Integer Programs" in Chapter 10.

Chapter 11 of this book introduces the practical application of linear and integer programming. Ultimately, real-world problems must be formulated as linear or integer programs and then solved on computers using commercial or public-domain software packages. We give examples and pointers to this end.

This book emphasizes intuitive concepts and gives a corresponding numerical example after each concept. It is intended to be a textbook for undergraduates and graduates, a short course, or self-learning. And, its unique approach draws on over 50 years of the first author's unique experience as researcher and educator, author of textbooks on combinatorial optimization and integer programming, and 40 years of teaching combinatorial optimization and algorithms to graduate and undergraduate students alike. The book's website, http://lipme.org, gives solutions to all exercises in the book, as well as additional exercises with solutions. The website also provides worked practical application examples and links to further resources.

The book was first typed by Alex Kahng and Aidan Kahng. It was retyped and proofread by Dr. Peng Du. Further valuable feedback and inputs were provided by Alex Kahng, along with Ilgweon Kang, Jiajia Li, Hyein Lee, Kwangsoo Han, Lutong Wang, Yaping Sun, and Mulong Luo. For their efforts, the authors are extremely grateful.

We hope that you will enjoy reading this book. Please write to us at lipme@vlsicad.ucsd.edu with your comments, your criticisms, or any errors that you have found. We will update you about changes in the second printing before it appears.

Last, the authors wish to dedicate this book to the following friends who have made important contributions to the contents of the book:

G.B. Dantzig
R.E. Gomory
A.J. Hoffman
E.L. Johnson
D.J. Kleitman
M.T. Shing

Otter Cove, CA T. C. Hu
La Jolla, CA A. B. Kahng
June 2015

Contents

1 Preliminaries

This chapter provides a brief linear algebra "refresher" and covers notation and some matrix properties that will be used throughout this book.

1.1 Matrix Properties

In this book, we will use parentheses to denote row vectors and brackets to denote column vectors. Thus, a matrix $\boldsymbol{A}$ defined as

$$\boldsymbol{A} = \begin{bmatrix} a_{11} & a_{12} & a_{13} & a_{14} & a_{15} \\ a_{21} & a_{22} & a_{23} & a_{24} & a_{25} \\ a_{31} & a_{32} & a_{33} & a_{34} & a_{35} \end{bmatrix}$$

can be written as three row vectors or five column vectors with $(a_{11}, a_{12}, a_{13}, a_{14}, a_{15})$ as its first row and $[a_{11}, a_{21}, a_{31}]$ as its first column. Notice that a_{ij} is the entry in the ith row and the jth column.

A matrix of m rows and n columns is denoted by $m \times n$ elements. Thus, for $m = 2$ and $n = 3$, we can have a matrix $\boldsymbol{A}$ defined as

$$\boldsymbol{A} = \begin{bmatrix} a_{11} & a_{12} & a_{13} \\ a_{21} & a_{22} & a_{23} \end{bmatrix}$$

or a matrix $\boldsymbol{B}$ defined as

$$\boldsymbol{B} = \begin{bmatrix} b_{11} & b_{12} & b_{13} \\ b_{21} & b_{22} & b_{23} \end{bmatrix}$$

T.C. Hu, A.B. Kahng, *Linear and Integer Programming Made Easy*,
DOI 10.1007/978-3-319-24001-5_1

We can also take the sum of these matrices to get

$$A + B = \begin{bmatrix} a_{11} + b_{11} & a_{12} + b_{12} & a_{13} + b_{13} \\ a_{21} + b_{21} & a_{22} + b_{22} & a_{23} + b_{23} \end{bmatrix}$$

In general, the sum of two $m \times n$ matrices is an $m \times n$ matrix whose entries are the sum of the corresponding entries of the summands.

A matrix $C = \begin{bmatrix} c_{11} & c_{12} \\ c_{21} & c_{22} \end{bmatrix}$ multiplied by a scalar α is $\alpha C = \begin{bmatrix} \alpha c_{11} & \alpha c_{12} \\ \alpha c_{21} & \alpha c_{22} \end{bmatrix}$.

In general, an $m \times n$ matrix C multiplied by a scalar α is the matrix obtained by multiplying each component of C by α.

The transpose of an $m \times n$ matrix A is denoted by A^{T}, which is an $n \times m$ matrix. The transpose of A is obtained by writing each row of A as a column instead. So, for matrices B and C below, we have B^{T} and C^{T} as

$$B = \begin{bmatrix} b_{11} & b_{12} & b_{13} \\ b_{21} & b_{22} & b_{23} \end{bmatrix} \quad B^{\mathrm{T}} = \begin{bmatrix} b_{11} & b_{21} \\ b_{12} & b_{22} \\ b_{13} & b_{23} \end{bmatrix}$$

$$C = \begin{bmatrix} c_{11} \\ c_{21} \\ c_{31} \end{bmatrix} \quad C^{\mathrm{T}} = \begin{bmatrix} c_{11} & c_{21} & c_{31} \end{bmatrix}$$

Now, let us look at some properties of square matrices. Consider an $n \times n$ square matrix C with

$$C = \begin{bmatrix} c_{11} & \cdots & c_{1n} \\ \vdots & \ddots & \vdots \\ c_{n1} & \cdots & c_{nn} \end{bmatrix}$$

We say C is the identity matrix I if $c_{ij} = 1$ if $i = j$ and $c_{ij} = 0$ if $i \neq j$ $(1 \leq i, j \leq n)$. Thus for a 3×3 square matrix, the identity matrix C is

$$C = \begin{bmatrix} 1 & 0 & 0 \\ 0 & 1 & 0 \\ 0 & 0 & 1 \end{bmatrix}$$

For a square matrix, we can define a determinant that is useful in linear algebra. The determinant of square matrix C is denoted as $\det C$. We will discuss this in Section 1.3.

Matrices can also be multiplied together. Two matrices A and B can be multiplied if the number of columns of A is the same as the number of rows of B. We could have

$$A = \begin{bmatrix} a_{11} & a_{12} & a_{13} \\ a_{21} & a_{22} & a_{23} \end{bmatrix}$$

$$B = \begin{bmatrix} b_{11} & b_{12} \\ b_{21} & b_{22} \\ b_{31} & b_{32} \end{bmatrix}$$

$$A \times B = \begin{bmatrix} a_{11}b_{11} + a_{12}b_{21} + a_{13}b_{31} & a_{11}b_{12} + a_{12}b_{22} + a_{13}b_{32} \\ a_{21}b_{11} + a_{22}b_{21} + a_{23}b_{31} & a_{21}b_{12} + a_{22}b_{22} + a_{23}b_{32} \end{bmatrix}$$

In general, an $m \times n$ matrix $\boldsymbol{A}$ and $n \times p$ matrix $\boldsymbol{B}$ can be multiplied to get an $m \times p$ matrix $\boldsymbol{C}$ where each element c_{ij} can be expressed as the dot product of the ith row of $\boldsymbol{A}$ and the jth column of $\boldsymbol{B}$.

We say that a matrix is in *row echelon form* if any rows containing all zeroes are below all rows with at least one nonzero element, and the leftmost nonzero element of each row is strictly to the right of the leftmost nonzero element of the row above it. So, the matrices $\boldsymbol{A}$ and $\boldsymbol{B}$ defined as

$$\boldsymbol{A} = \begin{bmatrix} 3 & 5 & 2 & 1 \\ 0 & 2 & 9 & 4 \\ 0 & 0 & 7 & 0 \end{bmatrix}, \quad \boldsymbol{B} = \begin{bmatrix} 1 & -4 & 1 & 6 \\ 0 & 0 & -3 & 3 \\ 0 & 0 & 0 & 0 \end{bmatrix}$$

are said to be in row echelon form.

We say that a matrix is in reduced row echelon form if it is in row echelon form, and the leftmost nonzero element of every row is 1 and all other elements in its column are zeroes. Thus, the matrices $\boldsymbol{C}$ and $\boldsymbol{D}$ defined as

$$\boldsymbol{C} = \begin{bmatrix} 1 & 0 & 0 & -3 \\ 0 & 1 & 0 & 2 \\ 0 & 0 & 1 & 0 \end{bmatrix}, \quad \boldsymbol{D} = \begin{bmatrix} 1 & -4 & 1 & 6 \\ 0 & 0 & 1 & -1 \\ 0 & 0 & 0 & 0 \end{bmatrix}$$

are said to be in reduced row echelon form. Note that the matrices $\boldsymbol{A}$ and $\boldsymbol{B}$ are not in reduced row echelon form.

1.2 Solving Simultaneous Equations

If we have a system of linear equations, we can express it as a matrix by just writing in the coefficients in what we call an augmented matrix. So, for the system

$$\begin{aligned} x_1 + 3x_2 + 2x_3 &= 7 \\ 0 + 2x_2 + x_3 &= 3 \\ x_1 - x_2 + x_3 &= 2 \end{aligned}$$

we have the augmented matrix

$$\begin{bmatrix} 1 & 3 & 2 & 7 \\ 0 & 2 & 1 & 3 \\ 1 & -1 & 1 & 2 \end{bmatrix}$$

This is just a convenient way for us to express and solve a system of linear equations.

When we solve a system of linear equations expressed as an augmented matrix, there are three operations called *elementary row operations* which we are allowed to perform on the rows. We can:

1. Swap two rows
2. Multiply a row by a scalar
3. Add a multiple of a row to another row

Notice that these operations all preserve the solution(s) to the system of linear equations. Swapping two rows clearly has no effect on the solutions, multiplying the right and left side of an equation does not affect its solutions, and adding two equations with common solutions preserves the solutions.

Gaussian Elimination

Gaussian elimination is a method used for simplifying a matrix. It operates as follows:

1. Swap rows so the leftmost column with a nonzero entry is nonzero in the first row.
2. Add multiples of the first row to every other row so that the entry in that column of every other row is zero.
3. Ignore the first row and repeat the process.

Please see the example below of the Gauss–Jordan Method to see how Gaussian elimination operates.

The Gauss–Jordan Method

The Gauss–Jordan Method first applies Gaussian elimination to get a row echelon form of the matrix. Then, the rows are reduced even further into reduced row echelon form. In the case of a square matrix, the reduced row echelon form is the identity matrix if the initial matrix was invertible.

To solve a set of n equations with n unknowns using the Gauss–Jordan Method, consider the example with $n = 3$ shown below:

$$\begin{aligned} x_1 + 3x_2 + 2x_3 &= 7 \\ 0 + 2x_2 + x_3 &= 3 \\ x_1 - x_2 + x_3 &= 2 \end{aligned}$$

We now illustrate the Gauss–Jordan Method:

$$\begin{aligned} x_1 + 3x_2 + 2x_3 &= 7 \\ 0 + 2x_2 + x_3 &= 3 \\ x_1 - x_2 + x_3 &= 2 \end{aligned} \qquad \begin{bmatrix} 1 & 3 & 2 & 7 \\ 0 & 2 & 1 & 3 \\ 1 & -1 & 1 & 2 \end{bmatrix}$$

Multiply row 1 by -1 and add it to row 3:

$$\begin{aligned} x_1 + 3x_2 + 2x_3 &= 7 \\ 0 + 2x_2 + x_3 &= 3 \\ 0 - 4x_2 - x_3 &= -5 \end{aligned} \qquad \begin{bmatrix} 1 & 3 & 2 & 7 \\ 0 & 2 & 1 & 3 \\ 0 & -4 & -1 & -5 \end{bmatrix}$$

Multiply row 2 by 2 and add it to row 3:

$$\begin{aligned} x_1 + 3x_2 + 2x_3 &= 7 \\ 0 + 2x_2 + x_3 &= 3 \\ 0 + 0 + x_3 &= 1 \end{aligned} \qquad \begin{bmatrix} 1 & 3 & 2 & 7 \\ 0 & 2 & 1 & 3 \\ 0 & 0 & 1 & 1 \end{bmatrix}$$

Notice that we now have an augmented matrix in row echelon form, and Gaussian elimination is complete. We now want to row reduce this matrix into reduced row echelon form.

Multiply row 2 by 0.5 to get a leading 1 in row 2:

$$\begin{aligned} x_1 + 3x_2 + 2x_3 &= 7 \\ 0 + x_2 + 0.5x_3 &= 1.5 \\ 0 + 0 + x_3 &= 1 \end{aligned} \qquad \begin{bmatrix} 1 & 3 & 2 & 7 \\ 0 & 1 & 0.5 & 1.5 \\ 0 & 0 & 1 & 1 \end{bmatrix}$$

Multiply row 2 by -3 and add it to row 1:

$$\begin{aligned} x_1 + 0 + 0.5x_3 &= 2.5 \\ 0 + x_2 + 0.5x_3 &= 1.5 \\ 0 + 0 + x_3 &= 1 \end{aligned} \qquad \begin{bmatrix} 1 & 0 & 0.5 & 2.5 \\ 0 & 1 & 0.5 & 1.5 \\ 0 & 0 & 1 & 1 \end{bmatrix}$$

Multiply row 3 by −0.5 and add it to each of rows 1 and 2:

$$\begin{aligned} x_1 + 0 + 0 &= 1 \\ 0 + x_2 + 0 &= 1 \\ 0 + 0 + x_3 &= 1 \end{aligned} \qquad \begin{bmatrix} 1 & 0 & 0 & 2 \\ 0 & 1 & 0 & 1 \\ 0 & 0 & 1 & 1 \end{bmatrix}$$

We thus have the solution $x_1 = 2$, $x_2 = x_3 = 1$.

1.3 Inverse of a Matrix

If the product $\boldsymbol{A} \times \boldsymbol{B}$ of two $n \times n$ square matrices $\boldsymbol{A}$ and $\boldsymbol{B}$ is the identity matrix $\boldsymbol{I}$, then we define $\boldsymbol{B}$ as the inverse matrix of $\boldsymbol{A}$ and denote $\boldsymbol{B} = \boldsymbol{A}^{-1}$. Likewise, we define $\boldsymbol{A}$ as the inverse matrix of $\boldsymbol{B}$ and denote $\boldsymbol{A} = \boldsymbol{B}^{-1}$.

If the product $\boldsymbol{C}_1 \times \boldsymbol{C}_2 = \boldsymbol{I}$, then $\boldsymbol{C}_1$ is called the left inverse matrix of $\boldsymbol{C}_2$, and $\boldsymbol{C}_2$ is called the right inverse matrix of $\boldsymbol{C}_1$.

For the special case of a 2 × 2 matrix $\boldsymbol{M} = \begin{bmatrix} \alpha & \beta \\ \gamma & \delta \end{bmatrix}$, there is an easy formula to find the inverse matrix of $\boldsymbol{M}$. Namely,

$$\boldsymbol{M}^{-1} = \begin{bmatrix} \alpha & \beta \\ \gamma & \delta \end{bmatrix}^{-1} = \frac{1}{\alpha\delta - \beta\gamma} \begin{bmatrix} \delta & -\beta \\ -\gamma & \alpha \end{bmatrix}$$

It is easy to verify that

$$\begin{bmatrix} \alpha & \beta \\ \gamma & \delta \end{bmatrix} \begin{bmatrix} \alpha & \beta \\ \gamma & \delta \end{bmatrix}^{-1} = \begin{bmatrix} 1 & 0 \\ 0 & 1 \end{bmatrix}$$

As an example, suppose we want to find the inverse of $\begin{bmatrix} 4 & 3 \\ 2 & 1 \end{bmatrix}$. We have

$$\begin{bmatrix} 4 & 3 \\ 2 & 1 \end{bmatrix}^{-1} = \frac{1}{-2} \begin{bmatrix} 1 & -3 \\ -2 & 4 \end{bmatrix}$$

It is left as an exercise for the reader to verify that this is really the inverse.

The determinant of the 2 × 2 square matrix $\boldsymbol{M} = \begin{bmatrix} \alpha & \beta \\ \gamma & \delta \end{bmatrix}$ is

$$\det \boldsymbol{M} = \alpha\delta - \beta\gamma$$

Note that if the determinant of a matrix is zero, then the matrix has no inverse.

For any $n \times n$ square matrix, there is an easy way to calculate its inverse matrix. First, we put an $n \times n$ identity matrix to the right of it. Then, we use Gaussian elimination to convert the left matrix into the identity matrix and execute the same row operations on the right matrix. Once the left matrix is the identity matrix, the matrix on the right will be the inverse of the initial matrix.

Accordingly, if we want to find the inverse of

$$\begin{bmatrix} 1 & 3 & 2 \\ 0 & 2 & 1 \\ 1 & -1 & 1 \end{bmatrix}$$

we put the identity matrix to the right to get

$$\begin{bmatrix} 1 & 3 & 2 \\ 0 & 2 & 1 \\ 1 & -1 & 1 \end{bmatrix} \begin{bmatrix} 1 & 0 & 0 \\ 0 & 1 & 0 \\ 0 & 0 & 1 \end{bmatrix}$$

Then we use Gaussian elimination to convert the left matrix into an identity matrix. We have

$$\begin{bmatrix} 1 & 3 & 2 \\ 0 & 2 & 1 \\ 1 & -1 & 1 \end{bmatrix} \begin{bmatrix} 1 & 0 & 0 \\ 0 & 1 & 0 \\ 0 & 0 & 1 \end{bmatrix} \quad \text{Multiply row 1 by } -1 \text{ and add it to row 3.}$$

$$\Leftrightarrow \begin{bmatrix} 1 & 3 & 2 \\ 0 & 2 & 1 \\ 0 & -4 & -1 \end{bmatrix} \begin{bmatrix} 1 & 0 & 0 \\ 0 & 1 & 0 \\ -1 & 0 & 1 \end{bmatrix} \quad \text{Multiply row 2 by } -1 \text{ and add it to row 1.}$$

$$\Leftrightarrow \begin{bmatrix} 1 & 1 & 1 \\ 0 & 2 & 1 \\ 0 & -4 & -1 \end{bmatrix} \begin{bmatrix} 1 & -1 & 0 \\ 0 & 1 & 0 \\ -1 & 0 & 1 \end{bmatrix} \quad \text{Multiply row 2 by 0.5 to obtain 1in the second row.}$$

$$\Leftrightarrow \begin{bmatrix} 1 & 1 & 1 \\ 0 & 1 & 0.5 \\ 0 & -4 & -1 \end{bmatrix} \begin{bmatrix} 1 & -1 & 0 \\ 0 & 0.5 & 0 \\ -1 & 0 & 1 \end{bmatrix} \quad \text{Multiply row 2 by } -1 \text{ and add it to row 1.}$$

$$\Leftrightarrow \begin{bmatrix} 1 & 0 & 0.5 \\ 0 & 1 & 0.5 \\ 0 & -4 & -1 \end{bmatrix} \begin{bmatrix} 1 & -1.5 & 0 \\ 0 & 0.5 & 0 \\ -1 & 0 & 1 \end{bmatrix} \quad \text{Multiply row 2 by 4 and add it to row 3.}$$

$$\Leftrightarrow \begin{bmatrix} 1 & 0 & 0.5 \\ 0 & 1 & 0.5 \\ 0 & 0 & 1 \end{bmatrix} \begin{bmatrix} 1 & -1.5 & 0 \\ 0 & 0.5 & 0 \\ -1 & 2 & 1 \end{bmatrix} \quad \text{Multiply row 3 by } -0.5 \text{ and add it to each of rows 1 and 2.}$$

$$\Leftrightarrow \begin{bmatrix} 1 & 0 & 0 \\ 0 & 1 & 0 \\ 0 & 0 & 1 \end{bmatrix} \begin{bmatrix} 1.5 & -2.5 & -0.5 \\ 0.5 & -0.5 & -0.5 \\ -1 & 2 & 1 \end{bmatrix}$$

We end up with

$$\begin{bmatrix} 1 & 3 & 2 \\ 0 & 2 & 1 \\ 1 & -1 & 1 \end{bmatrix}^{-1} = \begin{bmatrix} 1.5 & -2.5 & -0.5 \\ 0.5 & -0.5 & -0.5 \\ -1 & 2 & 1 \end{bmatrix}$$

As an exercise, the reader should verify that this is the inverse of the given matrix.

1.4 Matrix Multiplication

The multiplication of matrices is associative; that is, $(\boldsymbol{AB})\boldsymbol{C} = \boldsymbol{A}(\boldsymbol{BC})$. Consider all of the ways to evaluate the product of four matrices $\boldsymbol{A}$, $\boldsymbol{B}$, $\boldsymbol{C}$, and $\boldsymbol{D}$.

$$\boldsymbol{ABCD} = ((\boldsymbol{AB})\boldsymbol{C})\boldsymbol{D} = \boldsymbol{A}((\boldsymbol{BC})\boldsymbol{D}) = (\boldsymbol{AB})(\boldsymbol{CD}) = (\boldsymbol{A}(\boldsymbol{BC}))\boldsymbol{D} = \boldsymbol{A}(\boldsymbol{B}(\boldsymbol{CD}))$$

Let us suppose that:

1. $\boldsymbol{A}$ has 1000 rows and 20 columns.
2. $\boldsymbol{B}$ has 20 rows and 1 column.
3. $\boldsymbol{C}$ has 1 row and 50 columns.
4. $\boldsymbol{D}$ has 50 rows and 100 columns.

The number of multiplications needed to obtain the final result depends on the order of multiplications! Table 1.1 shows the number of multiplications to obtain the final result for each order of matrix multiplications.

Table 1.1 Number of multiplications to obtain the result with different orders of multiplication

$((\boldsymbol{AB})\boldsymbol{C})\boldsymbol{D}$	$(1000 \times 20 \times 1) + (1000 \times 1 \times 50) + (1000 \times 50 \times 100)$	$= 5,070,000$
$\boldsymbol{A}((\boldsymbol{BC})\boldsymbol{D})$	$(20 \times 1 \times 50) + (20 \times 50 \times 100) + (20 \times 100 \times 1000)$	$= 2,101,000$
$(\boldsymbol{AB})(\boldsymbol{CD})$	$(1 \times 20 \times 1000) + (1 \times 50 \times 100) + (1 \times 100 \times 1000)$	$= 125,000$
$(\boldsymbol{A}(\boldsymbol{BC}))\boldsymbol{D}$	$(50 \times 1 \times 20) + (50 \times 20 \times 1000) + (50 \times 100 \times 1000)$	$= 6,001,000$
$\boldsymbol{A}(\boldsymbol{B}(\boldsymbol{CD}))$	$(100 \times 50 \times 1) + (100 \times 1 \times 20) + (100 \times 1000 \times 20)$	$= 2,007,000$

The optimum order of multiplying a chain of matrices was studied since the 1970s. It was discovered by Hu and Shing [5][1] that the optimum order of multiplication of n matrices,

$$\boldsymbol{M}_1 \cdot \boldsymbol{M}_2 \cdot \boldsymbol{M}_3 \cdot \ldots \cdot \boldsymbol{M}_{n-1} = \boldsymbol{M}$$

is equivalent to optimizing the partitioning of an n-sided polygon into triangles, where the cost of a triangle is the product of the three numbers associated with the three vertices. Figure 1.1 shows the resulting view of multiplying four matrices with different orders.

[1] T. C. Hu and M. T. Shing, *Combinatorial Algorithms*, Dover, 2001

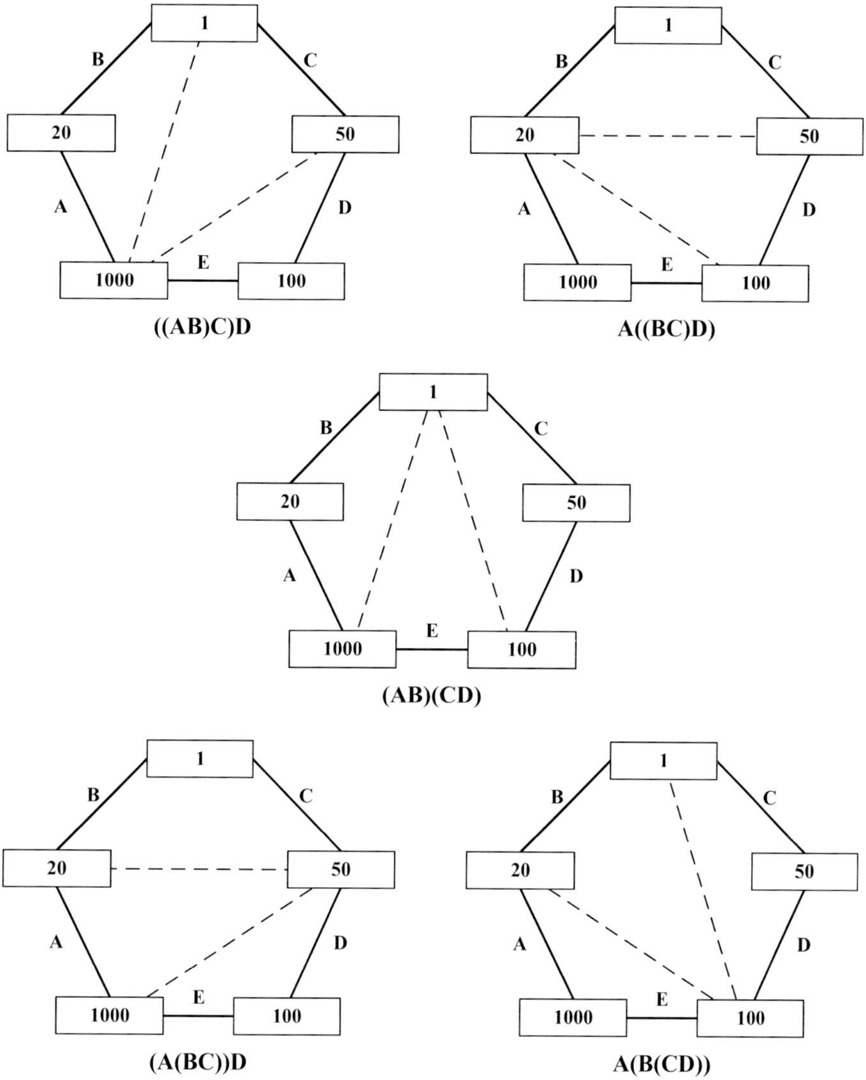

Fig. 1.1 The resulting view of multiplying four matrices with different orders

1.5 Exercises

1. Compute the following matrix products:

   (a) $\begin{bmatrix} 2 & 3 \\ 1 & 2 \end{bmatrix} \begin{bmatrix} 4 & 6 \\ 3 & 5 \end{bmatrix}$

(b) $\begin{bmatrix} 3 & 5 & -1 \\ -3 & 7 & 0 \end{bmatrix} \begin{bmatrix} 9 \\ -5 \\ 4 \end{bmatrix}$

2. Compute the inverse of each of the following matrices by placing the identity matrix to the right of it and using row reduction:

(a) $\begin{bmatrix} 4 & 3 \\ 3 & 2 \end{bmatrix}$

(b) $\begin{bmatrix} 7 & 2 & 1 \\ 0 & 3 & -1 \\ -3 & 4 & -2 \end{bmatrix}$

3. Compute the determinant of each of the following matrices:

(a) $\begin{bmatrix} 6 & 9 \\ 1 & -1 \end{bmatrix}$

(b) $\begin{bmatrix} 5 & 3 \\ 7 & 11 \end{bmatrix}$

4. What is the formula for finding the inverse of the square matrix $\begin{bmatrix} \propto & \beta \\ \gamma & \delta \end{bmatrix}$?

Apply it to find the inverse of

$$\begin{bmatrix} 7 & 3 \\ 2 & 1 \end{bmatrix}$$

5. Explain how Gaussian elimination can be used to solve a system of equations.
6. In a system of $n \geq 3$ simultaneous equations, would we get the final solution by multiplying any equation by a nonzero constant and adding it to another equation?
7. Solve the following systems of linear equations using the Gauss–Jordan Method:

(a) $\begin{cases} x + y + z = 5 \\ 2x + 3y + 5z = 8 \\ 4x + 5z = 2 \end{cases}$

(b) $\begin{cases} 4y + z = 2 \\ 2x + 6y - 2z = 3 \\ 4x + 8y - 5z = 4 \end{cases}$

(c) $\begin{cases} a + b + 2c = 1 \\ 2a - b + d = -2 \\ a - b - c - 2d = 4 \\ 2a - b + 2c - d = 0 \end{cases}$

8. How do you divide a 2×2 square matrix $\boldsymbol{A}$ by another square matrix $\boldsymbol{B}$?

9. Suppose that $\boldsymbol{A}$ is a 10×100 matrix, $\boldsymbol{B}$ is a 100×5 matrix, $\boldsymbol{C}$ is a 5×50 matrix, and $\boldsymbol{D}$ is a 50×1 matrix. Compute the number of multiplications needed in order to compute each of the following. Then, state which order of multiplication you found to be optimal.
   (a) $((\boldsymbol{AB})\boldsymbol{C})\boldsymbol{D}$
   (b) $(\boldsymbol{AB})(\boldsymbol{CD})$
   (c) $(\boldsymbol{A}(\boldsymbol{BC}))\boldsymbol{D}$
   (d) $\boldsymbol{A}((\boldsymbol{BC})\boldsymbol{D})$
   (e) $\boldsymbol{A}(\boldsymbol{B}(\boldsymbol{CD}))$

We reviewed some basic linear algebra in this chapter. Solutions to the exercises, as well as additional exercises, may be found at the book's website, http://lipme.org. Hopefully, you found the material to be easily accessible. From now on, we will be discussing linear and integer programming.

Introduction

2

We will now explore the basics of linear programming. In this chapter, we will go through several basic definitions and examples to build our intuition about linear programs. We will also learn the ratio method, which we will find useful in solving certain linear programs.

2.1 Linear Programming

Linear programming is a branch of mathematics just as calculus is a branch of mathematics. Calculus was developed in the seventeenth century to solve problems of physical sciences. Linear programming was developed in the twentieth century to solve problems of social sciences. Today, linear programming or its more generalized version, mathematical programming, has proven to be useful in many diversified fields: economics, management, all branches of engineering, physics, chemistry, and even pure mathematics itself. Linear programming can be viewed as a generalization of solving simultaneous linear equations. If solving a system of simultaneous linear equations can be considered a cornerstone of applied mathematics, then it is not surprising that linear programming has become so prominent in applied mathematics.

In the real world, we may want to maximize profits, minimize costs, maximize speed, minimize the area of a chip, etc. In maximizing profits, we are constrained by the limited resources available or the availability of the market. In minimizing the costs of production, we are constrained to satisfy certain standards. The idea of maximizing or minimizing a function subject to constraints arises naturally in many fields. In linear programming, we assume that the function to be maximized or minimized is a linear function and that the constraints are linear equations or linear inequalities. The assumptions of a linear model may not always be realistic, but it is the first approximate model for understanding a real-world problem. Let us now see the basic concept of a linear program.

T.C. Hu, A.B. Kahng, *Linear and Integer Programming Made Easy*,
DOI 10.1007/978-3-319-24001-5_2

Table 2.1 Simple example of a linear program

	Wood (units)	Iron (units)	Revenue	Output
Product A	12	6	$4	??
Product B	8	9	$5	??
Total resource	96	72	??	

Consider the following problem. We have two resources, wood and iron. We can use up to 96 units of wood per day, and we can use up to 72 units of iron per day. With these two resources, we can make two products, A and B. It requires 12 units of wood and 6 units of iron to produce one copy of product A, and it requires 8 units of wood and 9 units of iron to produce one copy of product B. Furthermore, we can sell products A and B for $4 and $5 per copy, respectively. How do we allocate resources to the production of A and B in order to maximize our revenue? Table 2.1 summarizes the problem.

With the information in Table 2.1, we can draw Figure 2.1 using A and B to represent the amounts of the two products. Two lines represent upper bounds on utilization of the two resources, wood and iron. For example, the line "wood" shows the maximum usage of wood. That is, the 96 available units of wood could potentially be used toward making eight copies of product A and zero copies of product B. Or, the wood could be used toward making 12 copies of product B and zero copies of product A. An important observation is that even though there is enough wood to make 12 copies of product B, this is not feasible since only 72 units of iron are available. The amounts of A and B produced must satisfy resource constraints on both wood and iron. As such, the polygon $JKLM$ represents our solution space, that is, the set of all feasible solutions to our problem.

Definition A *feasible solution* is a solution to a linear program that satisfies all constraints. The set of all feasible solutions is called the *solution space*. If a linear program has feasible solutions, then we say that it is *feasible*. Otherwise, it is said to be *infeasible*.

To maximize revenue, we have the objective to *maximize* $4A + 5B$. This means that we want to find a feasible solution that has maximum projection on the ray passing from the origin with a slope of $\frac{5}{4}$, shown as the vector $\boldsymbol{c}$ in Figure 2.2. We can draw a line with a slope of $-\frac{4}{5}$ (the dotted line in Figure 2.2), which is perpendicular to the vector $\boldsymbol{c}$. Shifting this dotted line along $\boldsymbol{c}$ allows us to find the feasible point with maximum projection. Figure 2.3 shows that the maximum possible revenue is $43.20, corresponding to production of 4.8 units each of product A and product B. However, if we are required to produce an integer number of units of each product, the integer solution (3, 6), which achieves revenue of $42.00, is optimum.

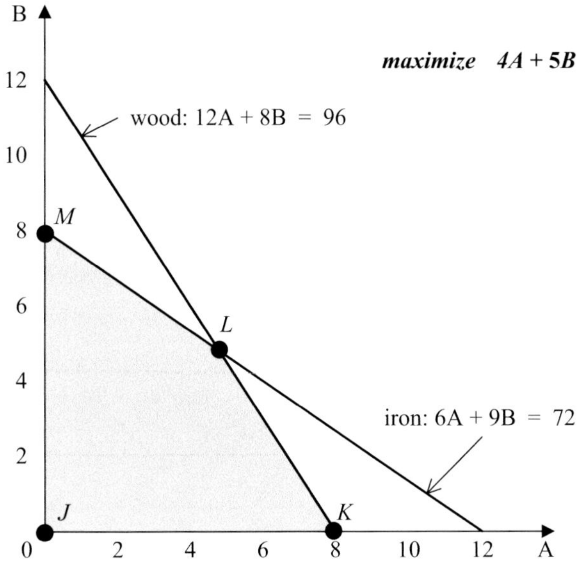

Fig. 2.1 Representation of the feasible regions of products A and B

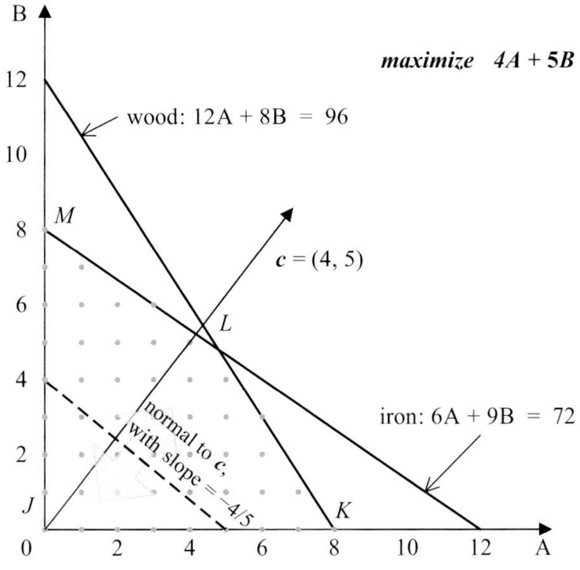

Fig. 2.2 Search for the feasible point with maximum projection by shifting the dotted line along $c(4,5)$

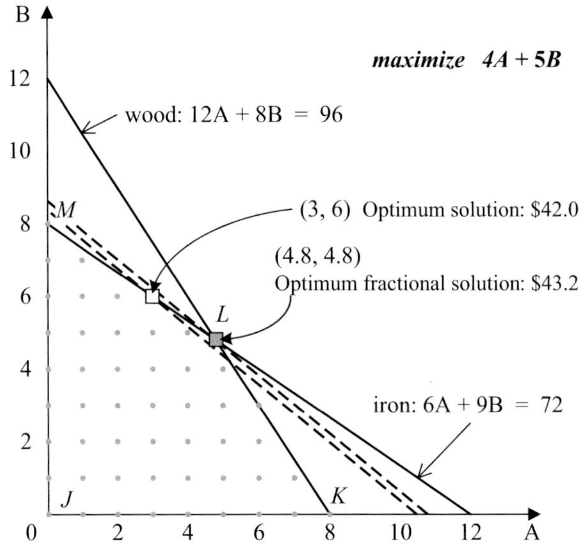

Fig. 2.3 Optimum solution and optimum fractional solution of the revenue

Let us consider one more linear program of two variables:

$$\begin{aligned} \max \quad & z = x_1 + x_2 \\ \text{subject to} \quad & 2x_1 + x_2 \leq 13 \\ & 0 \leq x_1 \leq 5 \\ & 0 \leq x_2 \leq 5 \end{aligned} \tag{2.1}$$

Notice that our solution space consists of the area bounded by *ABCDE* in Figure 2.4.

In Figure 2.4, the coordinates of the vertices and their values are $A = (0,\ 0)$ with value $z = 0$, $B = (5,\ 0)$ with value $z = 5$, $C = (5, 3)$ with value $z = 8$, $D = (4, 5)$ with value $z = 9$, and $E = (0, 5)$ with value $z = 5$.

The famous Simplex Method for solving a linear program—which we will present in Chapter 3—starts from vertex A, moves along the x_1 axis to B, and then moves vertically upward until it reaches C. From the vertex C, the Simplex Method goes upward and slightly westward to D. At the vertex D, the gradient of the function z increases toward the northeast, and the gradient is a convex combination of the vectors that are normal to lines ED and DC. This indicates that the corner point D is the vertex that maximizes the value of z[1] (Figure 2.5):

$$\max \quad z = \max(x_1 + x_2) = (4 + 5) = 9$$

[1] A *corner point* is also called an *extreme point*. Such a point is not "between" any other two points in the region.

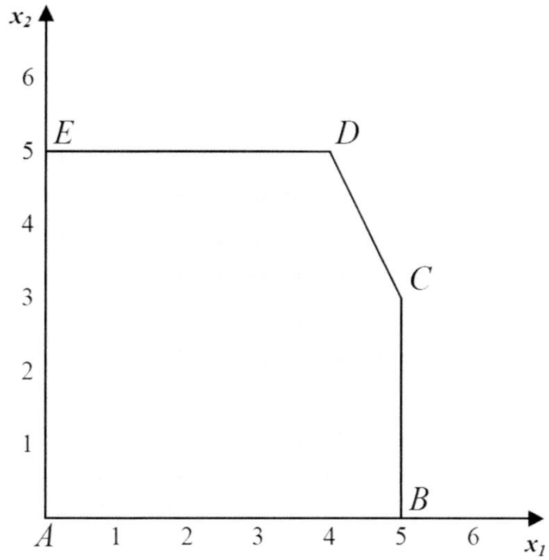

Fig. 2.4 Solution space of (2.1) is bounded by ABCDE

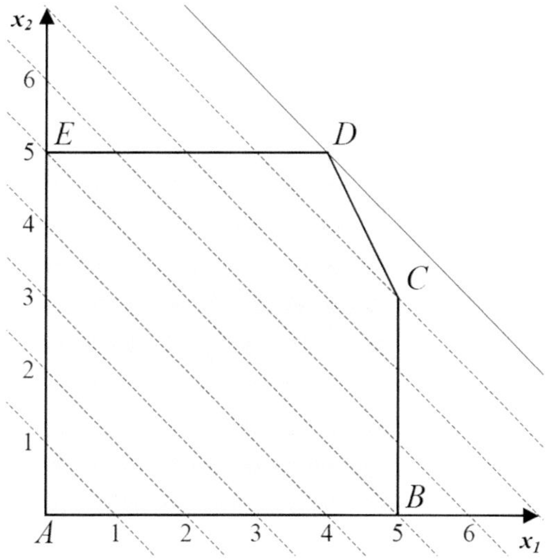

Fig. 2.5 Optimal solution of (2.1) is at the corner point D

For three variables x_1, x_2, and x_3, we can imagine a twisted cube such as a single die, where the minimum is at the lowest corner point and the maximum is at the highest corner point. Then, we can imagine that the Simplex Method would traverse along the boundary of the cube to search for the optimal solution among extreme points.

Definition A *linear program* is the maximization or minimization of a linear function subject to linear constraints (e.g., linear equations or linear inequalities).

For example, the following are two linear programs:

$$\begin{aligned} \max \quad & z = 12x_1 + 10x_2 + x_3 \\ \text{subject to} \quad & 11x_1 + 10x_2 + 9x_3 \le 20 \\ & x_1, x_2, x_3 \ge 0 \end{aligned} \tag{2.2}$$

$$\begin{aligned} \max \quad & z = x_1 + x_2 + x_3 \\ \text{subject to} \quad & 6x_1 + 3x_2 + x_3 \le 15 \\ & 4x_1 + 5x_2 + 6x_3 \le 15 \\ & x_1, x_2, x_3 \ge 0 \end{aligned} \tag{2.3}$$

There are three distinct features of linear programs:

1. The function to be maximized is a linear function.
2. The variables are restricted to be non-negative.
3. The constraints are linear inequalities or linear equations.

Thus, when solving a linear program, the classical tool of "calculus" is not used.

Definition The subject of optimization under constraints is called *mathematical programming*. An optimization problem with a function to be maximized or minimized is called a *mathematical program*.

Definition The function to be maximized or minimized in a mathematical program is called the *objective function*.

Definition If the objective function or the constraints are nonlinear in a mathematical program, then the mathematical program is a *nonlinear program*.

Definition If the variables are restricted to be non-negative integers in a mathematical program, then the mathematical program is called an *integer program*.

2.2 Ratio Method

In this section, we will familiarize ourselves with linear programs and the ratio method. We will proceed by exploring several examples.

Example 1 Consider a bread merchant carrying a knapsack who goes to the farmers market. He can fill his knapsack with three kinds of bread to sell. A loaf of raisin bread can be sold for \$12, a loaf of wheat bread for \$10, and a loaf of white bread for \$1.

Furthermore, the loaves of raisin, wheat, and white bread weigh 11 lbs, 10 lbs, and 9 lbs, respectively. If the bread merchant can carry 20 lbs of bread in the knapsack, what kinds of breads should he carry if he wants to get the most possible cash?

This problem can be formulated as a linear program (2.2) where x_1 denotes the number of loaves of raisin bread, x_2 denotes the number of loaves of wheat bread, and x_3 denotes the number of loaves of white bread in his knapsack.

In this example, there are three "noteworthy" integer solutions: $x_1 = x_3 = 1$, $x_2 = 0$ or $x_1 = x_3 = 0, x_2 = 2$ or $x_1 = x_2 = 0, x_3 = 2$. So, the merchant would carry two loaves of wheat bread and get \$20, rather than one loaf each of raisin bread and white bread to get \$13 or two loaves of white bread to get \$2. However, notice that if he can cut a loaf of raisin bread into pieces, $\frac{9}{11}$ of a loaf of raisin bread is worth $\$12 \times \frac{9}{11} \approx \$9.82$, and he should therefore carry $\frac{20}{11} = 1 + \frac{9}{11}$ loaves of raisin bread with market value of $\$21.82 = \$12 + \$9.82$.

In a linear program, the variables are not required to be integers, so it is much easier to solve than an integer program which requires the variables to be integers. Let us consider a linear program with only a single constraint. We have

$$\begin{aligned} \max \quad & z = v_1x_1 + v_2x_2 + v_3x_3 \\ \text{subject to} \quad & w_1x_1 + w_2x_2 + w_3x_3 \le b \\ & x_1, x_2, x_3 \ge 0 \end{aligned} \tag{2.4}$$

where v_j is the value associated with item j, w_j is the weight associated with item j, and b is the total weight restriction of the knapsack.

Intuitively, we would like to carry an item which is of low weight and high value. In other words, the ratio of the value to weight for such an item should be maximized. Applying this idea to (2.2), we have

$$\frac{12}{11} > \frac{10}{10} > \frac{1}{9}$$
$$\text{(i.e., raisin bread} > \text{wheat bread} > \text{white bread)}$$

which indicates that we should fill the knapsack with the first item, that is, carry $\frac{20}{11} = 1 + \frac{9}{11}$ loaves of raisin bread.

Definition The *ratio method* is a method that can be applied to a linear program to get the optimal solution as long as the variables are not restricted to be integers. It operates simply by taking the maximum or minimum ratio between two appropriate parameters (e.g., value to cost, or value to weight).

To formalize this idea of the ratio method, consider the general linear program

$$\begin{aligned} \max \quad & v = v_1x_1 + v_2x_2 + v_3x_3 + \cdots \\ \text{subject to} \quad & w_1x_1 + w_2x_2 + w_3x_3 + \cdots \le b \\ & x_j \ge 0 \end{aligned} \tag{2.5}$$

Furthermore, let $v_k/w_k = \max_j\{v_j/w_j\}$ (where $w_j \neq 0$). Then the feasible solution that maximizes v is $x_k = b/w_k, x_j = 0$ for $j \neq k$, and the maximum profit value of v is obtained by filling the knapsack with a single item. In total, we obtain a profit of $v = b \cdot (v_k/w_k)$ dollars.

It is easy to prove that the max ratio method for selecting the best item is correct for any number of items and any right-hand side (total weight restriction). It should be emphasized that the variables in (2.5) are not restricted to be integers. Normally, when we say a "knapsack problem," we refer to a problem like (2.2) but with integer constraints. To distinguish the difference, we call (2.2) a "fractional knapsack problem."

Example 2 Consider another merchant who goes to the farmers market. His goal is not to get the maximum amount of cash but to minimize the total weight of his knapsack as long as he can receive enough money. He also has the choice of three kinds of bread: raisin, wheat, and white (see (2.6)). Then his problem becomes minimizing the total weight subject to the amount of cash received being at least, say, \$30, as shown in (2.7).

$$\begin{aligned} \min \quad & z = w_1x_1 + w_2x_2 + w_3x_3 \\ \text{subject to} \quad & v_1x_1 + v_2x_2 + v_3x_3 \geq c \\ & x_j \geq 0 \end{aligned} \tag{2.6}$$

where v_j is the value associated with item j, w_j is the weight associated with item j, and c is the minimum amount of cash received.

When the objective function is to minimize the total weight, we also take ratios associated with the items, but we want the minimum ratio of weight to value. We have

$$\begin{aligned} \min \quad & z = 11x_1 + 10x_2 + 9x_3 \\ \text{subject to} \quad & 12x_1 + 10x_2 + 1x_3 \geq 30 \\ & x_j \geq 0 \end{aligned} \tag{2.7}$$

and the minimum ratio is $\frac{11}{12} < \frac{10}{10} < \frac{9}{1}$, which means that we should carry $\frac{30}{12} = 2.5$ loaves of the raisin bread, and the total weight of the knapsack is 11 lbs $\times$ 2.5 = 27.5 lbs.

Thus, in a fractional knapsack problem, we always take a ratio involving weight and value. To maximize profit, we use the maximum ratio of value to weight. To minimize weight, we use the reciprocal, that is, the minimum ratio of weight to value. The ratio method is always correct for a single constraint and no integer restrictions.

Example 3 Consider a merchant selling three kinds of drinks, A, B, and C. The three kinds of drinks are all made by mixing two kinds of juices, apple juice and orange juice. All drinks are sold at the market for one dollar per gallon. The merchant has 15 gallons of apple juice and 15 gallons of orange juice. To mix the juices to form drink A, the ratio of apple juice to orange juice is 4:1. For B, the ratio is 1:4, and for C, the ratio is 3:2. He has to mix the drinks before going to the market, but he is able to carry any combination of the three drinks. To get the maximum amount of cash, he must solve a problem equivalent to (2.8):

$$\begin{aligned} \max \quad & z = x_1 + x_2 + x_3 \\ \text{subject to} \quad & \begin{bmatrix} 4 \\ 1 \end{bmatrix} x_1 + \begin{bmatrix} 1 \\ 4 \end{bmatrix} x_2 + \begin{bmatrix} 3 \\ 2 \end{bmatrix} x_3 \leq \begin{bmatrix} 15 \\ 15 \end{bmatrix} \\ & x_j \geq 0 \end{aligned} \tag{2.8}$$

where x_1, x_2, and x_3 are the amounts of drinks A, B, and C, respectively.

Definition Mixing the apple juice and orange juice into drink A in a specified proportion is an *activity*. That is, an activity is a column vector in the above linear program, e.g., $\begin{bmatrix} 4 \\ 1 \end{bmatrix}, \begin{bmatrix} 1 \\ 4 \end{bmatrix}$, or $\begin{bmatrix} 3 \\ 2 \end{bmatrix}$. Also in the example above, the amount of a drink that is produced is called the *activity level*, e.g., x_1 or x_2.

If the activity level is zero, i.e., $x_1 = 0$, it means that we do not select that activity at all. In this example, we have three activities, so we have three activity levels to choose. (In fact, the book of T. C. Koopmans (ed.) in 1951 is called *Activity Analysis of Production and Allocation*.)

Here, x_1, x_2, and x_3 are the amounts of drinks A, B, and C, respectively, that the merchant will carry to the market. By trial and error, we might find that an optimum solution is $x_1 = 3$, $x_2 = 3$, and $x_3 = 0$, or $[x_1, x_2, x_3] = [3, 3, 0]$.

Note that every drink needs five parts of juice, and if we take the ratio of value to weight as before, all the ratios are equal. We have

$$\frac{1}{4+1} = \frac{1}{1+4} = \frac{1}{3+2} = \frac{1}{5} \tag{2.9}$$

The reason that we do not select drinks A and C or B and C in the optimum solution is that drinks A and B are more compatible in the sense that they do not compete for the same kinds of juices heavily.

Definition Two activities are *compatible* if the right-hand side (RHS) of the constraint function can be expressed as a sum of the two activities with non-negative activity level coefficients.

The notion of compatibility is central to a linear program with more than one constraint. To illustrate this point, we let the ratio of apple juice to orange juice for drink A be 2:3. We keep the ratios for drink B and drink C the same. Thus, we have the linear program

$$\begin{aligned} \max \quad & v = x_1 + x_2 + x_3 \\ \text{subject to} \quad & \begin{bmatrix} 2 \\ 3 \end{bmatrix} x_1 + \begin{bmatrix} 1 \\ 4 \end{bmatrix} x_2 + \begin{bmatrix} 3 \\ 2 \end{bmatrix} x_3 \leq \begin{bmatrix} 15 \\ 15 \end{bmatrix} \\ & x_j \geq 0 \end{aligned} \tag{2.10}$$

Now, the optimum solution is $x_1 = x_3 = 3$ and $x_2 = 0$. Note that we have neither changed the prices of any drinks nor the composition of drinks B and C. Now, drink B is not selected and drink C is selected. The reason is that now drinks A and B are not compatible, whereas drinks A and C are compatible.

If you are the coach of a basketball team, you do not want a team of players who are all only good at shooting. You need a variety of players who can also pass and run fast. The same need for ability in diverse areas is necessary to create compatible teams in most sports.

Similarly, when it comes to linear programming, we are not trying to find just the best column vector, but we want to find the best set of vectors. Just like in the real world, we are not seeking the best single player but the best team of players.

We shall write a linear program of n variables and m constraints as

$$\begin{aligned} \max \quad & v = \sum c_j x_j \\ \text{subject to} \quad & \sum a_{ij} x_j \leq b_i \quad (i = 1, \ldots, m)(j = 1, \ldots, n) \\ & x_j \geq 0 \end{aligned} \tag{2.11}$$

(Strictly speaking, we should count the inequalities $x_j \geq 0$ as constraints, but in most applications, we understand that the requirement of non-negativity of variables is treated separately.)

In matrix notation, we write

$$\begin{aligned} \max \quad & v = \boldsymbol{cx} \\ \text{subject to} \quad & \boldsymbol{Ax} \leq \boldsymbol{b} \\ & \boldsymbol{x} \geq \boldsymbol{0} \end{aligned} \tag{2.12}$$

where $\boldsymbol{c}$ is a row vector which has c_j as its component $(j = 1, \ldots, n)$, $\boldsymbol{x}$ is a column vector which has x_j as its component $(j = 1, \ldots, n)$, $\boldsymbol{b}$ is a column vector which has b_i as its component $(i = 1, \ldots, m)$, $\boldsymbol{0}$ is a zero vector, and the matrix $\boldsymbol{A}$ is an $m \times n$ matrix with its columns denoted by $\boldsymbol{a}_1, \boldsymbol{a}_2, \ldots, \boldsymbol{a}_n$.

One of the earliest applications of linear programming was to find the optimum selection of food items within a budget, i.e., to discover what kinds of food items satisfy the nutritional requirements while minimizing the total cost. We call this problem the Homemaker Problem.

For the purposes of discussion, let us assume that we are concerned with the nutritional requirements for vitamins A and B only. As such, we shall represent every food in the supermarket as a vector with two components, the first component being the amount of vitamin A the food contains and the second component being the amount of vitamin B it contains. For example, we shall represent beef as [3, 1], meaning that a pound of beef contains three units of vitamin A and one unit of vitamin B. Similarly, we may represent wheat as [1, 1]. We may also potentially represent a food as [−1, 2], meaning that that particular food will destroy one unit of vitamin A but provide two units of vitamin B.

Thus, the process of deciding to buy a particular food is equivalent to selecting a column vector $\boldsymbol{a}_j$ in the matrix $\boldsymbol{A}$. There are n column vectors, so we say that there are n activities. The amount of a particular food j to be purchased is called its activity level and is denoted by x_j. For example, if we associate j with beef, then $x_j = 3$ means that we should buy three pounds of beef, and $x_j = 0$ means that we should not buy any beef. Since we can only buy from the supermarket, it is natural to require $x_j \geq 0$. The unit cost of food j is denoted by c_j, so the total bill for all purchases is $\sum c_j x_j$. The total amount of vitamin A in all the foodstuffs purchased is $\sum a_{1j} x_j$, and similarly, the amount of vitamin B is $\sum a_{2j} x_j$. As such, the linear program describing our problem is

$$\begin{aligned} \min \quad & z = \sum c_j x_j \\ \text{subject to} \quad & \sum a_{ij} x_j \geq b_i \quad (i = 1, \ldots, m)(j = 1, \ldots, n) \\ & x_j \geq 0 \end{aligned} \tag{2.13}$$

Example 4 *(Homemaker Problem)* Assume that the supermarket stocks four kinds of food costing \$15, \$7, \$4, and \$6 per pound, and that the relevant nutritional characteristics of the food can be represented as

$$\begin{bmatrix} 3 \\ 1 \end{bmatrix}, \quad \begin{bmatrix} 1 \\ 1 \end{bmatrix}, \quad \begin{bmatrix} 0 \\ 1 \end{bmatrix}, \quad \begin{bmatrix} -1 \\ 2 \end{bmatrix}.$$

If we know the nutritional requirements for vitamins A and B for the whole family are [3, 5], then we have the following linear program:

$$\begin{aligned} \min \quad & z = 15x_1 + 7x_2 + 4x_3 + 6x_4 \\ \text{subject to} \quad & \begin{bmatrix} 3 \\ 1 \end{bmatrix} x_1 + \begin{bmatrix} 1 \\ 1 \end{bmatrix} x_2 + \begin{bmatrix} 0 \\ 1 \end{bmatrix} x_3 + \begin{bmatrix} -1 \\ 2 \end{bmatrix} x_4 \geq \begin{bmatrix} 3 \\ 5 \end{bmatrix} \\ & x_j \geq 0 \quad (j = 1, 2, 3, 4) \end{aligned} \tag{2.14}$$

The optimum solution turns out to be $x_1 = 0$, $x_2 = 3$, $x_3 = 2$, and $x_4 = 0$. This means that we should buy three pounds of the second food and two pounds of the third food, and none of the first or the fourth food.

Example 5 *(Pill Salesperson)* Let us twist the Homemaker Problem a little bit and consider a vitamin pill salesperson who wants to compete with the supermarket. Since the taste and other properties of the food are of no concern (by assumption), the salesperson merely wants to provide pills that contain equivalent nutrition at a lower cost than the food. Let us assume that there are two kinds of pill, one for vitamin A and one for vitamin B, and each pill supplies one unit of its vitamin. Suppose that the salesperson sets the price of vitamin A pills at y_1 and vitamin B pills at y_2. If the prices satisfy the constraints

$$\begin{aligned} 3y_1 + \ y_2 &\leq 15 \\ y_1 + \ y_2 &\leq 7 \\ y_2 &\leq 4 \\ -y_1 + 2y_2 &\leq 6 \end{aligned}$$

then no matter what combination of food items we select from the supermarket, it is always cheaper to satisfy our nutritional requirements by buying the pills. Since the requirements for vitamins A and B are [3, 5], the total amount that we have to pay the salesperson is $3y_1 + 5y_2$. Of course, in setting prices, the salesperson would like to maximize the amount he receives for his goods. Once again, we can represent his problem as a linear program:

$$\begin{aligned} \max \quad & z = 3y_1 + 5y_2 \\ \text{subject to} \quad & 3y_1 + \ y_2 \leq 15 \\ & y_1 + \ y_2 \leq 7 \\ & y_2 \leq 4 \\ & -y_1 + 2y_2 \leq 6 \\ & y_1, y_2 \geq 0 \end{aligned} \tag{2.15}$$

Now, let us study a general method to solve the linear program with a single constraint. The following are two linear programs, each of which has a single constraint:

$$\begin{aligned} \min \quad & z = x_1 + 2x_2 + 3x_3 \\ \text{subject to} \quad & 4x_1 + 5x_2 + 6x_3 \geq 60 \\ & x_j \geq 0 \quad (j = 1, 2, 3) \end{aligned} \tag{2.16}$$

$$\begin{aligned} \max \quad & v = x_1 + 2x_2 + 3x_3 \\ \text{subject to} \quad & 6x_1 + 5x_2 + 4x_3 \leq 60 \\ & x_j \geq 0 \quad (j = 1, 2, 3) \end{aligned} \tag{2.17}$$

To solve the minimization problem, let us take one variable at a time. To satisfy (3.15) using x_1, we need

$$x_1 = \frac{60}{4} = 15 \quad \text{and} \quad z = \$15$$

$$\text{For } x_2, \; x_2 = \frac{60}{5} = 12 \quad \text{and} \quad z = \$24$$

$$\text{For } x_3, \; x_3 = \frac{60}{6} = 10 \quad \text{and} \quad z = \$30$$

The intuitive idea is to take the ratios $\left(\frac{1}{4}\right)$, $\left(\frac{2}{5}\right)$, and $\left(\frac{3}{6}\right)$ and select the minimum ratio if we want to minimize. If we want to maximize, the ratios should be $\left(\frac{1}{6}\right)$, $\left(\frac{2}{5}\right)$, and $\left(\frac{3}{4}\right)$, and we select the maximum ratio.

$$\text{For } x_1, \; x_1 = \frac{60}{6} = 10 \quad \text{and} \quad v = \$10$$

$$\text{For } x_2, \; x_2 = \frac{60}{5} = 12 \quad \text{and} \quad v = \$24$$

$$\text{For } x_3, \; x_3 = \frac{60}{4} = 15 \quad \text{and} \quad v = \$45$$

To formalize the idea, let the linear program be

$$\begin{aligned} \min \quad & z = c_1x_1 + c_2x_2 + \cdots + c_nx_n \\ \text{subject to} \quad & a_1x_1 + a_2x_2 + \cdots + a_nx_n \geq b \\ & x_j \geq 0 \quad (j = 1, \ldots, n) \end{aligned} \tag{2.18}$$

Then we select the minimum ratio

$$\min_j \left(\frac{c_j}{a_j}\right) = \frac{c_k}{a_k} \quad (\text{where } a_j \neq 0)$$

and let $x_k = \frac{b}{a_k}$ and $x_j = 0$ if $j \neq k$.

If the linear program is

$$\begin{aligned} \max \quad & v = c_1x_1 + c_2x_2 + \cdots + c_nx_n \\ \text{subject to} \quad & a_1x_1 + a_2x_2 + \cdots + a_nx_n \leq b \\ & x_j \geq 0 \end{aligned} \tag{2.19}$$

then we select the maximum ratio

$$\max_j \left(\frac{c_j}{a_j}\right) = \frac{c_k}{a_k} \quad (\text{where } a_j \neq 0)$$

and let $x_k = \frac{b}{a_k}$, and $x_j = 0$ if $j \neq k$.

This is a formal restatement of the ratio method.

Congratulations! We have just developed a method to solve any linear program with a single constraint and any number of variables!

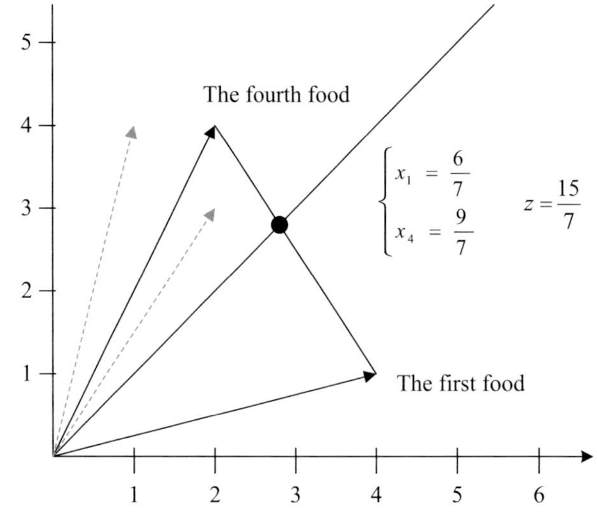

Fig. 2.6 Graphic solution of the Homemaker Problem described in (2.20)

Let us now consider the Homemaker Problem in Example 4 again, but with a different set of data. Assume that the supermarket has four kinds of food, each kind labeled by its vitamin contents of vitamin A and vitamin B, as follows:

$$\begin{bmatrix} 4 \\ 1 \end{bmatrix}, \quad \begin{bmatrix} 1 \\ 4 \end{bmatrix}, \quad \begin{bmatrix} 3 \\ 2 \end{bmatrix}, \quad \begin{bmatrix} 4 \\ 8 \end{bmatrix}.$$

All items cost one dollar per pound, except the fourth item which costs two dollars per pound. And the homemaker needs 6 units of vitamin A and 6 units of vitamin B for his family. So the homemaker's linear program becomes

$$\begin{aligned} \min \quad & z = x_1 + x_2 + x_3 + x_4 \\ \text{subject to} \quad & \begin{bmatrix} 4 \\ 1 \end{bmatrix} x_1 + \begin{bmatrix} 1 \\ 4 \end{bmatrix} x_2 + \begin{bmatrix} 3 \\ 2 \end{bmatrix} x_3 + \begin{bmatrix} 2 \\ 4 \end{bmatrix} x_4 \geq \begin{bmatrix} 6 \\ 6 \end{bmatrix} \\ & x_j \geq 0 \quad (j = 1, 2, 3, 4) \end{aligned} \tag{2.20}$$

Note that the vector $\begin{bmatrix} 4 \\ 8 \end{bmatrix}$ has been normalized to $\begin{bmatrix} 2 \\ 4 \end{bmatrix}$ and costs one dollar. Note also that (2.20) has two constraints, and we want to solve the problem by a graphic method. In Figure 2.6, the horizontal axis measures the amount of vitamin A and the vertical axis measures the amount of vitamin B of the item. Also, there is a line from the origin to $\begin{bmatrix} 6 \\ 6 \end{bmatrix}$.

Since we have two constraints, we need to select the best pair of items. The pair which intersects the 45° line furthest from the origin is the cheapest pair. For (2.20), $x_1 = \frac{6}{7}$, $x_4 = \frac{9}{7}$, and $z = \frac{15}{7}$.

The Pill Salesperson Problem corresponding to (2.20) becomes (2.21) where the vitamin A pill costs π_1 dollars and the vitamin B pill costs π_2 dollars:

$$\begin{aligned} \max \quad & v = 6\pi_1 + 6\pi_2 \\ \text{subject to} \quad & 4\pi_1 + \pi_2 \le 1 \\ & \pi_1 + 4\pi_2 \le 1 \\ & 3\pi_1 + 2\pi_2 \le 1 \\ & 2\pi_1 + 4\pi_2 \le 1 \end{aligned} \tag{2.21}$$

The optimum solution is $\pi_1 = \frac{3}{14}$, $\pi_2 = \frac{2}{14}$, and $v = \frac{15}{7}$, same as the value of $z = \frac{15}{7}$. This can be found graphically by finding the intersection point of $4\pi_1 + \pi_2 = 1$ and $2\pi_1 + 4\pi_2 = 1$, where $\pi_1 = \frac{3}{14}$ and $\pi_2 = \frac{2}{14}$.

2.3 Exercises

1. Draw the solution space graphically for the following problems. Then, determine the optimum fractional solutions using your graph.
   (a) Producing one copy of A requires 3 units of water and 4 units of electricity, while producing one copy of B requires 7 units of water and 6 units of electricity. The profits of products A and B are 2 units and 4 units, respectively. Assuming that the factory can use up to 21 units of water and 24 units of electricity per day, how should the resources be allocated in order to maximize profits?
   (b) Suppose there are two brands of vitamin pills. One pill of Brand A costs \$3 and contains 30 units of vitamin X and 10 units of vitamin Y. One pill of Brand B costs \$4 and contains 15 units of vitamin X and 15 units of vitamin Y. If you need 60 units of vitamin X and 30 units of vitamin E and want to minimize your spending, how will you purchase your pills?
2. Why do we use $\max x_0$,

$$x_0 - c_1x_1 - c_2x_2 - \cdots - c_nx_n = 0,$$

   and use $\min z$,

$$-z + c_1x_1 + c_2x_2 + \cdots + c_nx_n = 0?$$

3. Use the ratio method to solve the following:
   (a)

$$\begin{aligned} \min \quad & z = x_1 + 3x_2 + 5x_3 \\ \text{subject to} \quad & 4x_1 + 7x_2 + 8x_3 \ge 120 \\ & x_j \ge 0 \end{aligned}$$

   (b)

$$\begin{aligned} \max \quad & z = 3x_1 + 8x_2 + x_3 \\ \text{subject to} \quad & 6x_1 + 15x_2 + 3x_3 \le 90 \\ & x_j \ge 0 \end{aligned}$$

4. Write linear programs for the following two problems.
   (a) A group of 25 adults and 18 children are going to travel. There are two types of vans they can rent. The first type accommodates six adults and six children and costs \$90 to rent for the duration of the trip. The second type accommodates six adults and four children, and it costs \$80. In order to minimize the cost, how should the group rent vans?
   (b) A school wants to create a meal for its students by mixing food A and food B. Each ounce of A contains 2 units of protein, 4 units of carbohydrates, and 2 units of fat. Each ounce of B contains 3 units of protein, 1 unit of carbohydrates, and 4 units of fat. If the meal must provide at least 10 units of protein and no more than 8 units of carbohydrates, how should the school create the meal in order to minimize the amount of fat?
5. Use the ratio method to solve

$$\begin{array}{rl} \min & z = 16x_1 + 9x_2 + 3x_3 \\ \text{subject to} & 8x_1 + 4x_2 + x_3 = 120 \\ & x_j \geq 0 \end{array}$$

6. Use the ratio method to solve

$$\begin{array}{rl} \min & z = 16x_1 + 7x_2 + 3x_3 \\ \text{subject to} & 8x_1 + 4x_2 + x_3 = 120 \\ & x_j \geq 0 \end{array}$$

7. In Exercises 5 and 6, both constraints are equations. Will the solutions change if the constraints are inequalities?
8. Prove that the ratio method works correctly for both maximizing and minimizing the objective function.

This chapter covered the basics of linear programming. You should now be comfortable with the concept of a linear program, how the ratio method works, and the notation that we have been using in order to express linear programs. It is highly advisable for you to be comfortable with the material in this chapter before moving on.

Also, you may have noticed that the Homemaker Problem and the Pill Salesperson Problem in the last section are closely related. It turns out that they are *dual* linear programs. We will be revisiting these problems later in Chapter 4 as we explore duality theory, and you will see why they both yielded the same optimal value for the objective function.

In the next chapter, we will discuss solution spaces and some properties of linear programs. Let's go!

Dimension of the Solution Space 3

In this chapter, we will discuss solution spaces, convex sets and convex functions, the geometry of linear programs, and the theorem of separating hyperplanes. We will apply what we cover in this chapter to build our understanding of the Simplex Method, duality theory, and other concepts in subsequent chapters.

3.1 Solution Spaces

Definition We call the number of coordinates needed to express any point in a space the *dimension* of the space. Thus, for example, you can think of a point as having zero dimensions, a line as having one dimension, and a plane (or a part of a plane) as having two dimensions.

Before we try to solve a linear program, let us first solve a simple set of simultaneous equations:

$$\begin{aligned} x_1 + x_2 &= 5 \\ x_1 - x_2 &= 1 \end{aligned} \tag{3.1}$$

Using Gaussian elimination, we find the equivalent simultaneous equations:

$$\begin{aligned} x_1 &= 3 \\ x_2 &= 2 \end{aligned} \tag{3.2}$$

In the two-dimensional plane, each of the equations $x_1 + x_2 = 5$ and $x_1 - x_2 = 1$ in (3.1) defines a line. Their intersection is the point $x_1 = 3$ and $x_2 = 2$ as shown in Figure 3.1. Since a point is of zero dimensions, we say that the solution space of (3.1) is of zero dimensions. In (3.2), $x_1 = 3$ defines a line and $x_2 = 2$ also defines a line, so again their intersection is the same point $x_1 = 3$, $x_2 = 2$, as shown by the dashed lines in Figure 3.1.

T.C. Hu, A.B. Kahng, *Linear and Integer Programming Made Easy*,
DOI 10.1007/978-3-319-24001-5_3

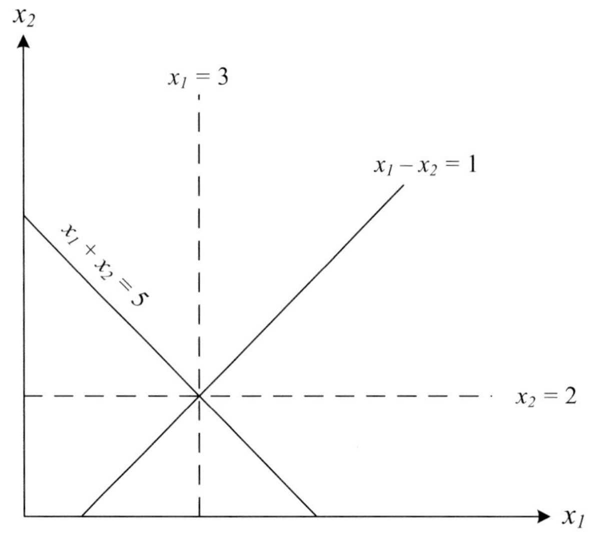

Fig. 3.1 Lines and their intersections

On the other hand, if there is only a single equation

$$x_1 + x_2 = 5$$

then there are infinitely many solutions:

$$\begin{bmatrix} x_1 \\ x_2 \end{bmatrix} = \begin{bmatrix} 5 \\ 0 \end{bmatrix}, \begin{bmatrix} 6 \\ -1 \end{bmatrix}, \begin{bmatrix} 4.5 \\ 0.5 \end{bmatrix}, \begin{bmatrix} 2.5 \\ 2.5 \end{bmatrix}, \text{ etc.}$$

In order to fix a point on the line $x_1 + x_2 = 5$, we need to specify one parameter, say $x_1 = 3.8$. Then the point is uniquely determined as $[x_1, x_2] = [3.8, 1.2]$. Hence, because we need to specify one parameter, we say that the line $x_1 + x_2 = 5$ is of one dimension. In a three-dimensional space, the equation $x_1 + 2x_2 + 3x_3 = 6$ defines a plane, and that plane is of two dimensions. In order to fix a point on the plane, we need to specify two parameters, say $x_1 = 2$ and $x_2 = 1$, in which case the third parameter and hence the point is uniquely determined:

$$x_3 = \frac{6 - x_1 - 2x_2}{3} = \frac{6 - 2 - 2(1)}{3} = \frac{2}{3}$$

Similarly, the circle $x_1^2 + x_2^2 = 25$ is of one dimension since we need to specify only one parameter, the angle from the x_1 axis, to determine a point uniquely.

Definition A system of linear equations is *consistent* if it has at least one solution (i.e., the solution space contains at least one point). Otherwise, it is *inconsistent*.

Definition A system of linear equations is *redundant* if one of the equations can be expressed as a linear combination of other equations. Otherwise, it is *nonredundant*.

In general, the solution space of m simultaneous linear equations in n variables is of $n - m$ dimensions if the equations are consistent and nonredundant. However, when the constraint is an inequality, the inequality does not reduce the dimensionality of the solution space. For example, the $x_1 - x_2$ plane is of two dimensions, since we need to fix two parameters to specify a point. If we constrain the space by the three inequalities

$$\begin{aligned} x_1 + x_2 &\leq 5 \\ x_1 &\geq 0 \\ x_2 &\geq 0 \end{aligned}$$

then the solution space is a triangular portion of the plane as we can see in Figure 3.1 and still of two dimensions.

We can now see that each inequality constraint defines a half-space, and the solution space is the intersection of all the half-spaces. We have also discussed the dimension of a solution space of a linear program and now want to talk about the shape of a solution space. It turns out the solution space always corresponds to a *convex set*.

If we consider a two-dimensional convex set as the area of a room, then a person can be anywhere inside the room, and he can see all other points in the convex set. Our formal definition of a convex set will capture this idea. In Figure 3.2 the top three shapes represent convex sets, while the bottom three do not.

3.2 Convex Sets and Convex Functions

Definition A set X is *convex* if for any two points x_1 and x_2 in the set, all points lying on the line segment connecting x_1 and x_2 also belong to the set. In other words, all points of the form

$$\lambda x_1 + (1 - \lambda)x_2$$

with $0 \leq \lambda \leq 1$ also belong to X.

For example, all points in the triangle defined by

$$\begin{aligned} x_1 + x_2 &\leq 5 \\ x_1 &\geq 0 \\ x_2 &\geq 0 \end{aligned}$$

form a convex set. All points on the $x_1 - x_2$ plane also form a convex set.

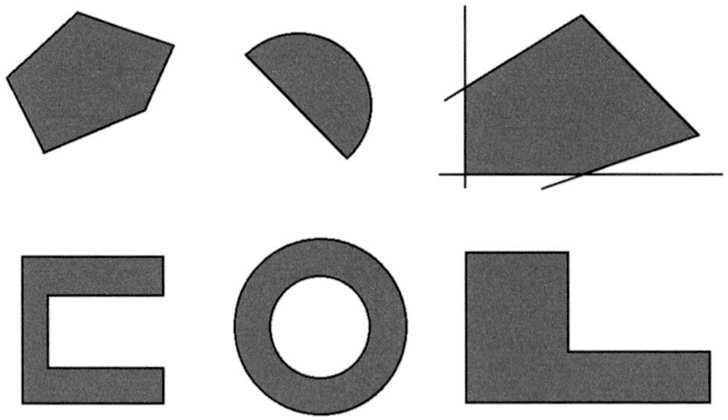

Fig. 3.2 Examples of convex (*top row*) and non-convex (*bottom row*) sets

Lemma 3.1 *The intersection of convex sets is convex.*

Definition A point x of a convex set X is an *extreme point* if and only if

$$x = \lambda x_1 + (1 - \lambda)x_2 \quad (0 < \lambda < 1)$$

implies $x_1 = x_2 = x$, or $x_1 \notin X$, or $x_2 \notin X$. In other words, there is no way to express an extreme point x of convex set X as an interior point of a line segment where the two distinct endpoints x_1 and x_2 both belong to X.

Definition A function $f(x)$ defined on a convex set X is a *convex function* if

$$f(\lambda x_1 + (1 - \lambda)x_2) \leq \lambda f(x_1) + (1 - \lambda)f(x_2) \quad (\text{for } 0 \leq \lambda \leq 1)$$

for any $x_1, x_2 \in X$. A convex function is a *strictly convex function* if

$$f(\lambda x_1 + (1 - \lambda)x_2) < \lambda f(x_1) + (1 - \lambda)f(x_2) \quad (\text{for } 0 < \lambda < 1)$$

for any two distinct points $x_1, x_2 \in X$.

Note that a convex function is always defined on a convex set X. Otherwise, the point $\lambda x_1 + (1 - \lambda)x_2$ may not be in X. Geometrically, if we consider X as a plane and $f(x)$ as a surface plotted above the plane, then the surface of the convex function has the property that a line segment connecting any two points on the surface lies entirely above or on the surface. For a strictly convex function, the line segment lies entirely above the surface except at the two endpoints.

Theorem 3.1 *If $f(x)$ is a convex function defined on a closed and bounded convex set X, then a local minimum (strict or not strict) of $f(x)$ is a global minimum of $f(x)$.*

Proof Let $f(x)$ have a local minimum at x_0, i.e., there exists a neighborhood of x_0 such that $f(x) \geq f(x_0)$ for $|x - x_0| < \epsilon$. Let x^* be any other point in X with $f(x^*) < f(x_0)$. All the points of the form $\lambda x^* + (1 - \lambda)x_0$ $(1 \geq \lambda \geq 0)$ belong to X since X is convex. We can take λ sufficiently small, e.g., ϵ^2, such that $\bar{x} = \lambda x^* + (1 - \lambda)x_0$ is in the ϵ-neighborhood of x_0. By the assumption that x_0 is a local minimum of $f(x)$, we have $f(\bar{x}) \geq f(x_0)$ or

$$\begin{aligned} f(\bar{x}) \geq f(x_0) &= \lambda f(x_0) + (1 - \lambda)f(x_0) \\ &> \lambda f(x^*) + (1 - \lambda)f(x_0) \quad (0 < \lambda \leq \epsilon^2) \end{aligned} \tag{3.3}$$

From the definition of convex functions,

$$f(\bar{x}) \leq \lambda f(x^*) + (1 - \lambda)f(x_0) \quad (0 \leq \lambda \leq 1) \tag{3.4}$$

Since $0 < \lambda \leq \epsilon^2$ is a subinterval of $0 \leq \lambda \leq 1$, (3.3) and (3.4) provide the desired contradiction. Note that we did not assume that there is a global minimum at x^*. The existence of a global minimum at x^* in a closed and bounded set requires the theorem of Weierstrass. We take x^* to be any point with $f(x^*) < f(x_0)$. □

Since a local minimum of a convex function implies a global minimum, we can use a greedy approach to find the minimum of a convex function.

Definition A *greedy algorithm* is an algorithm that iteratively makes a locally optimum choice (e.g., following the steepest gradient) in an attempt to find a globally optimum solution.

Definition The negative or opposite of a convex function is called a *concave function*. Thus, a function $f(x)$ defined on a convex set is called a concave function if

$$f(\lambda x_1 + (1 - \lambda)x_2) \geq \lambda f(x_1) + (1 - \lambda)f(x_2) \quad (\text{for } 0 \leq \lambda \leq 1)$$

for any $x_1, x_2 \in X$. If the inequality in the above definition is replaced by a strict inequality, then the function is called a *strictly concave function*.

Theorem 3.2 *If $f(x)$ is a concave function defined on a closed and bounded convex set X, then a global minimum of $f(x)$ exists at an extreme point of X.*

Proof Let v_i be the extreme points of X and let $x = \sum \lambda_i v_i$ $\left(\sum \lambda_i = 1, \lambda_i \geq 0\right)$. Then $f(x) = f\left(\sum \lambda_i v_i\right) \geq \sum \lambda_i f(v_i) \geq \min_i f(v_i)$. □

A linear function is both a convex function and a concave function. Thus, for a linear function, a local minimum is a global minimum, and a global minimum occurs at an extreme point.

3.3 Geometric Interpretation of Linear Programs

Consider a linear program

$$\begin{aligned} \max \quad & z = \boldsymbol{cx} \\ \text{subject to} \quad & \boldsymbol{Ax} \leq \boldsymbol{b} \\ & \boldsymbol{x} \geq \boldsymbol{0} \end{aligned} \tag{3.5}$$

where $\boldsymbol{A}$ is an $m \times n$ matrix, $\boldsymbol{x}$ is a column vector of n components, $\boldsymbol{b}$ is a column vector of m components, and $\boldsymbol{c}$ is a row vector of n components.

Definition The m inequalities together with the non-negativity constraints of $\boldsymbol{x}$ define a convex set in the space of $\boldsymbol{x}$ called the *activity space*. In other words, the activity space is the convex set consisting of every point that satisfies all constraints of the linear program.

Every point in the convex set (also known as a convex *polytope*) is a feasible solution of the linear program. Among all feasible solutions, we want the point which maximizes the objective function $z = \boldsymbol{cx}$. Consider a set of parallel hyperplanes:

$$\begin{aligned} -3x_1 + 2x_2 + 4x_3 &= 16 \\ -3x_1 + 2x_2 + 4x_3 &= 14 \\ -3x_1 + 2x_2 + 4x_3 &= 12 \end{aligned}$$

Each hyperplane has the same normal vector $(-3, 2, 4)$. We can imagine a hyperplane with normal vector $(-3, 2, 4)$ moving along its normal vector in $x_1 - x_2 - x_3$ space; for each position, there is an associated value of a parameter k such as 16, 14, 12, etc. At certain positions, the hyperplane intersects the convex set defined by the constraints of the linear program.

If we imagine that the hyperplane moves from smaller values of the parameter k to larger values, then there will be an intersection point of the convex set with the hyperplane of largest value, which is the optimum solution of the linear program.

In general, the optimum solution of a linear program is a vertex of the convex polytope (see Figure 3.3a). However, if the hyperplane defined by the objective function $z = \boldsymbol{cx}$ is parallel to one of the faces of the convex polytope, then every point on the face is an optimum solution (see Figure 3.3b).

Consider the linear program

$$\begin{aligned} \min \quad & z = \sum c_j x_j \quad (j = 1, \ldots, n) \\ \text{subject to} \quad & \sum a_{ij} x_j = b_i \quad (i = 1, \ldots, m) \\ & x_j \geq 0 \end{aligned} \tag{3.6}$$

In most cases, n is much larger than m. Let us assume that $n = 100$ and $m = 2$ so that the activity space is of $100 - 2 = 98$ dimensions. There is another way of

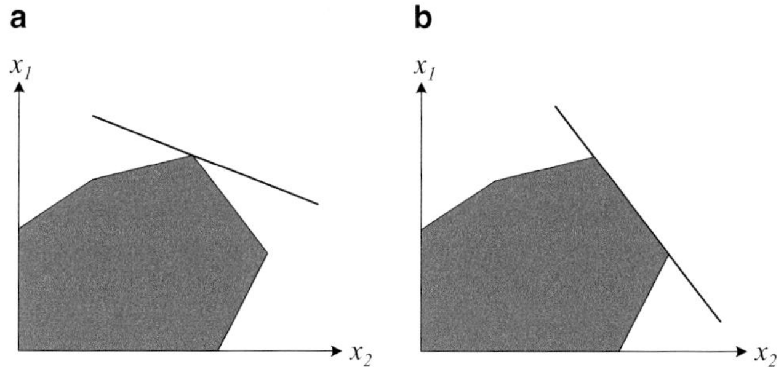

Fig. 3.3 The optimal solution of a linear program: (**a**) a vertex of a convex polytope; and (**b**) every point on the face

viewing the linear program geometrically. Namely, we have 100 column vectors, each having two components. In the Homemaker Problem, the 100 column vectors represent 100 kinds of food, each containing different amounts of vitamins A and B. The nutrition requirements of the homemaker's family are also represented by a two-component vector $\boldsymbol{b}$ with components b_1, the amount of vitamin A, and b_2, the amount of vitamin B.

Under the linear program model, we can buy half a loaf of bread for one dollar if the whole loaf is priced at two dollars and one third of a watermelon for one dollar if the whole watermelon is three dollars. For simplicity, we can let each column vector have two components which represent the amount of vitamins A and B in one dollar's worth of that kind of food.

Let us now set up a coordinate system where the horizontal axis denotes the amount of vitamin A contained in an item and the vertical axis denotes the amount of vitamin B contained in an item. We can plot the 100 items as 100 vectors where the length of a vector is determined by what one dollar can buy. The nutrition requirement vector $\boldsymbol{b}$ can also be plotted on the same diagram. This is shown in Figure 3.4 where only five vectors instead of 100 vectors are plotted.

Note that these vectors follow the rule of vector addition. If $\boldsymbol{a}_1 = [a_{11}, a_{21}]$ and $\boldsymbol{a}_2 = [a_{12}, a_{22}]$, then $\boldsymbol{a}_1 + \boldsymbol{a}_2 = \boldsymbol{a}_3$ where $\boldsymbol{a}_3$ has the components $[a_{11} + a_{12},\ a_{21} + a_{22}]$. In this two-dimensional space, we want to select two vectors of minimum total cost, with each component of their vector sum greater than or equal to the corresponding component of $\boldsymbol{b}$.

3.4 Theorem of Separating Hyperplanes

Theorem 3.3 (Theorem of Separating Hyperplanes) *There are three equivalent forms of the theorem of separating hyperplanes*:

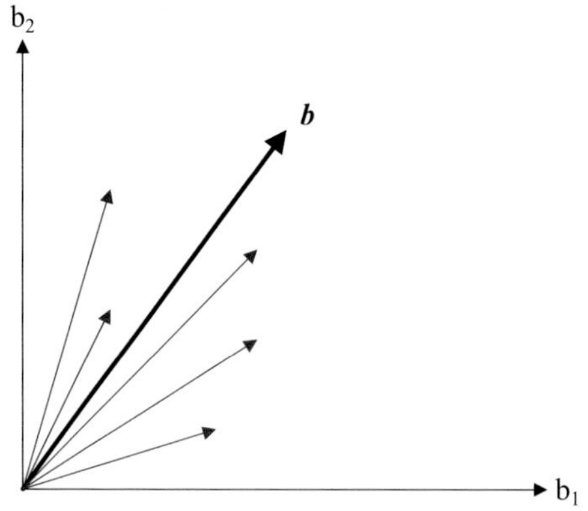

Fig. 3.4 Vectors representing nutrition requirements

1. *Either the inequality* $\boldsymbol{Ax} \geq \boldsymbol{b}$ *has a non-negative solution* $\boldsymbol{x}$ *or the inequalities*

$$\boldsymbol{yA} \leq \boldsymbol{0}$$
$$\boldsymbol{yb} > \boldsymbol{0}$$

have a non-negative solution $\boldsymbol{y}$.

2. *Either the inequality* $\boldsymbol{Ax} \leq \boldsymbol{b}$ *has a non-negative solution* $\boldsymbol{x}$ *or the inequalities*

$$\boldsymbol{yA} \geq \boldsymbol{0}$$
$$\boldsymbol{yb} < \boldsymbol{0}$$

have a non-negative solution $\boldsymbol{y}$.

3. *Either the equation* $\boldsymbol{Ax} = \boldsymbol{b}$ $(\boldsymbol{b} \neq \boldsymbol{0})$ *has a non-negative solution* $\boldsymbol{x}$ *or the inequalities*

$$\boldsymbol{yA} \geq \boldsymbol{0}$$
$$\boldsymbol{yb} < \boldsymbol{0}$$

have a solution $\boldsymbol{y}$ *(not necessarily non-negative).*

For the case of (1), geometrically, if $\boldsymbol{Ax} \geq \boldsymbol{b}$ has a non-negative solution, then the vector $\boldsymbol{b}$ lies within the cone generated (i.e., spanned) by the columns of $\boldsymbol{A}$. If there is no non-negative solution, then there exists a hyperplane separating all columns of $\boldsymbol{A}$ on one side and a vector $\boldsymbol{y}$ on the other side. The vector $\boldsymbol{y}$ should make an angle of less than 90° with $\boldsymbol{b}$, which is perpendicular to the hyperplane. The cases (2) and (3) have similar interpretations.

For a numerical example, let us assume

$$\boldsymbol{A} = \begin{bmatrix} -2 & -3 \\ 4 & 3 \end{bmatrix}, \quad \boldsymbol{b} = \begin{bmatrix} -4 \\ 1 \end{bmatrix}$$

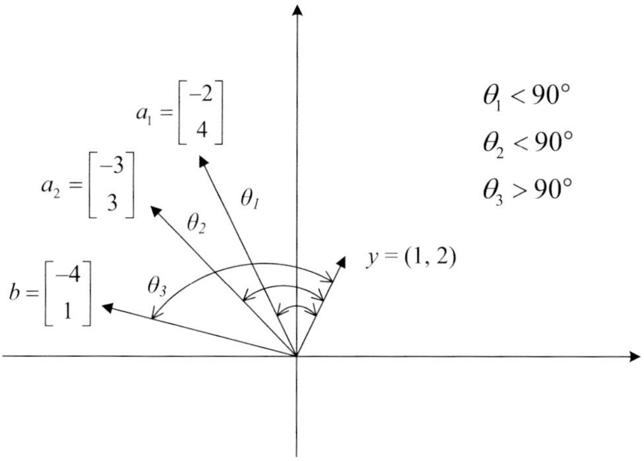

Fig. 3.5 Illustration of case 3 for the Theorem of Separating Hyperplanes

Consider the equation

$$\begin{bmatrix} -2 & -3 \\ 4 & 3 \end{bmatrix} \begin{bmatrix} x_1 \\ x_2 \end{bmatrix} = \begin{bmatrix} -4 \\ 1 \end{bmatrix}$$

for which there is no non-negative solution $\boldsymbol{x}$. However, consider $\boldsymbol{yb}$ for the vector $\boldsymbol{y} = (1, 2)$, as shown in Figure 3.5,

$$[1 \quad 2] \begin{bmatrix} -2 \\ 4 \end{bmatrix} = 6 \geq 0$$

$$[1 \quad 2] \begin{bmatrix} -3 \\ 3 \end{bmatrix} = 3 \geq 0$$

$$[1 \quad 2] \begin{bmatrix} -4 \\ 1 \end{bmatrix} = -2 < 0$$

3.5 Exercises

1. What is a linear program? What is a constraint of a linear program? Does it make a difference if the constraint is an equation or an inequality?
2. Given a linear program with seven variables, two equality constraints, and three inequality constraints, what is the dimension of the solution space of the linear program?

3. Is the following system of linear equations consistent? Why or why not?

$$\begin{aligned} x_1 \quad\quad + x_3 &= 1 \\ x_1 + x_2 + x_3 &= 2 \\ x_1 - x_2 + x_3 &= 1 \end{aligned}$$

4. Is the following system of linear equations redundant? Why or why not?

$$\begin{aligned} 3x_1 + \ x_2 - 6x_3 &= -10 \\ 2x_1 + \ x_2 - 5x_3 &= -8 \\ 6x_1 - 3x_2 + 3x_3 &= 0 \end{aligned}$$

5. Is the following system of linear equations consistent and redundant? Why or why not?

$$\begin{aligned} 2x_1 + \ x_2 + \ 3x_3 &= 1 \\ 2x_1 + 6x_2 + \ 8x_3 &= 3 \\ 6x_1 + 8x_2 + 18x_3 &= 5 \end{aligned}$$

6. We say two linear programs are equivalent if they have the same unique optimum solutions and the dimensions of their solution spaces are the same. Are the linear programs in (3.7)–(3.9) equivalent? Why or why not?

$$\begin{aligned} \text{max} \quad & v = x_1 + 2x_2 + 3x_3 \\ \text{subject to} \quad & 12x_1 + 12x_2 + \ 6x_3 \leq 30 \\ & \ 4x_1 + 10x_2 + 18x_3 \leq 15 \\ & x_j \geq 0 \end{aligned} \tag{3.7}$$

First divide the first inequality by 2, yielding this modified problem:

$$\begin{aligned} \text{max} \quad & v = x_1 + 2x_2 + 3x_3 \\ \text{subject to} \quad & 6x_1 + \ 6x_2 + \ 3x_3 \leq 15 \\ & 4x_1 + 10x_2 + 18x_3 \leq 15 \\ & x_j \geq 0 \end{aligned} \tag{3.8}$$

Then divide the second column by 2 and the third column by 3.

$$\begin{aligned} \text{max} \quad & v = x_1 + x_2 + x_3 \\ \text{subject to} \quad & 6x_1 + 3x_2 + \ x_3 \leq 15 \\ & 4x_1 + 5x_2 + 6x_3 \leq 15 \\ & x_j \geq 0 \end{aligned} \tag{3.9}$$

Good job on completing this chapter! We now have all the knowledge needed before learning about fundamental linear programming techniques such as the Simplex Method—in Chapter 4, next!

4 Introduction to the Simplex Method

In this chapter, we will learn the Simplex Method, which is a widely used technique for solving linear programs. Even though the notation can be a bit daunting, the technique is actually quite simple. Just keep in mind that the Simplex Method essentially involves iterative application of the *ratio test* (which you already know) and the *pivot operation* (which we are about to cover).

4.1 Equivalent Formulations

Geometrically, it is easy to view an inequality as a constraint. Computationally, it is much easier to work with equations. As such, it is convenient for us to introduce the idea of a slack variable.

Definition A *slack variable* is an introduced variable used to convert a constraint in the form of an inequality to an equation.

Here, we shall summarize some standard tricks for converting between equivalent formulations. We have already seen some conversions between inequalities and equations, and we now systematically list all of these. More conversions that help in formulating real-world problems as linear programs are given as "tips and tricks" in Section 11.2 and on this book's website.

1. *An inequality can be converted into an equation by introducing a slack variable.* For instance, the inequality

$$x_1 + 2x_2 + 3x_3 \leq 10$$

T.C. Hu, A.B. Kahng, *Linear and Integer Programming Made Easy*,
DOI 10.1007/978-3-319-24001-5_4

is equivalent to

$$x_1 + 2x_2 + 3x_3 + s_1 = 10$$
$$s_1 \geq 0 \quad (s_1 \text{ is a slack variable})$$

Similarly, the inequality

$$5x_4 + 6x_5 - 3x_6 \geq 12$$

is equivalent to

$$5x_4 + 6x_5 - 3x_6 - s_2 = 12$$
$$s_2 \geq 0 \quad (s_2 \text{ is a slack variable})$$

2. A single equation can be replaced by two inequalities, and in general, *m equations can be replaced by* $m + 1$ *inequalities*. For example, the equation $3x_1 + 2x_2 + x_3 = 10$ can be replaced by

$$3x_1 + 2x_2 + x_3 \leq 10$$
$$3x_1 + 2x_2 + x_3 \geq 10$$

Similarly, any system of m equations

$$\sum_j a_{ij}x_j = b_i \quad (i = 1, \ldots, m) \; (j = 1, \ldots, n)$$

can be replaced by the $m + 1$ inequalities

$$\sum_j a_{ij}x_j \geq b_i \quad (i = 1, \ldots, m) \; (j = 1, \ldots, n)$$
$$\sum_i \left(\sum_j a_{ij}x_j - b_i \right) \leq 0$$

3. *Maximizing a function is the same as minimizing the negative of the function.* For example,

$$\max \; x_1 - 2x_2 + 3x_3 - 4x_4$$

is the same as

$$\min \; -x_1 + 2x_2 - 3x_3 + 4x_4$$

Using these conversions, we can put a linear program into an equivalent form. The reader should convince himself or herself that these conversions are valid and that the dimension of the solution space remains the same for all three conversions. If you are interested, please see Chapter 11 for additional tricks in formulating problems as linear programs!

4.2 Basic, Feasible, and Optimum Solutions

From now on, we shall always assume that the system of linear equations is consistent and nonredundant. Recall that this means a system of linear equations has at least one solution and that none of the equations can be expressed as a linear combination of the other equations. Also, by convention, we will assume that there are m rows (equations) and n columns (variables) in the matrix, and all variables are required to be non-negative.

Definition A set of vectors is *linearly independent* if none of the vectors can be expressed as a linear combination of the others. For example, the three vectors in the plane

$$\begin{bmatrix} 4 \\ 5 \end{bmatrix}, \begin{bmatrix} -1 \\ 2 \end{bmatrix}, \begin{bmatrix} 2 \\ 9 \end{bmatrix}$$

are not linearly independent since

$$\begin{bmatrix} 4 \\ 5 \end{bmatrix} + 2\begin{bmatrix} -1 \\ 2 \end{bmatrix} = \begin{bmatrix} 2 \\ 9 \end{bmatrix}$$

On the other hand, the three vectors

$$\begin{bmatrix} 4 \\ 0 \\ 6 \end{bmatrix}, \begin{bmatrix} 0 \\ 4 \\ 2 \end{bmatrix}, \begin{bmatrix} 2 \\ 1 \\ 1 \end{bmatrix}$$

are linearly independent.

In three-dimensional space, any vector can be expressed as a linear combination of

$$\begin{bmatrix} 1 \\ 0 \\ 0 \end{bmatrix}, \begin{bmatrix} 0 \\ 1 \\ 0 \end{bmatrix}, \begin{bmatrix} 0 \\ 0 \\ 1 \end{bmatrix}$$

or any other set of three linearly independent vectors.

Definition A linear program has an *unbounded solution* if the solution can be made arbitrarily large while satisfying all constraints.

Definition A variable x_j is an *unrestricted variable* if there are no restrictions on the sign of x_j, i.e., we cannot define x_j as either $x_j \geq 0$ or $x_j \leq 0$.

For the following definitions, let $\boldsymbol{A} = [\boldsymbol{B}, \boldsymbol{N}]$ and $\boldsymbol{x} = [\boldsymbol{x}_B, \boldsymbol{x}_N]$, where $\boldsymbol{B}$ is a nonsingular (i.e., invertible) square matrix, $\boldsymbol{x}_B$ is a vector with m components, and

$\boldsymbol{x}_N$ is a vector with $n-m$ components. Also, we let $\boldsymbol{b}$ be a vector with m components.

Definition The solution of $\boldsymbol{Ax}=\boldsymbol{b}$ obtained by setting $\boldsymbol{x}_B=\boldsymbol{B}^{-1}\boldsymbol{b}$ and $\boldsymbol{x}_N=\boldsymbol{0}$ is called a *basic solution*. We call the components of $\boldsymbol{x}_B$ the *basic variables* and the components of $\boldsymbol{x}_N$ the *non-basic variables*.

Definition A basic solution is called *degenerate* if one or more components of $\boldsymbol{x}_B$ are zero.

Definition The column vectors of $\boldsymbol{B}$ are called the *basic vectors*. Sometimes we say that the m independent column vectors of $\boldsymbol{B}$ form the *basis*.

Definition A feasible solution is called a *basic feasible solution* if the set of column vectors corresponding to positive components of the solution is independent.

For brevity, we shall say "solution" for a feasible solution and "basic solution" for a basic feasible solution.

Consider the following system of equations.

$$\begin{bmatrix}4\\8\\10\end{bmatrix}x_1+\begin{bmatrix}1\\0\\0\end{bmatrix}x_2+\begin{bmatrix}0\\2\\1\end{bmatrix}x_3+\begin{bmatrix}0\\0\\3\end{bmatrix}x_4=\begin{bmatrix}3\\4\\5\end{bmatrix},\quad x_j\geq 0\quad (j=1,2,3,4)$$

- $[x_1,x_2,x_3,x_4]=[0,3,2,1]$ is a basic solution.
- $[x_1,x_2,x_3,x_4]=\left[\frac{1}{2},1,0,0\right]$ is a degenerate basic solution.
- $[x_1,x_2,x_3,x_4]=\left[\frac{1}{4},2,1,\frac{1}{2}\right]$ is a feasible solution but is not basic.

Theorem 4.1 *If there exists a solution to a system of linear equations, there exists a basic solution to the linear equations.*

A solution to linear equations is a *feasible solution* if all of its components are non-negative. We shall now prove an important extension to Theorem 4.1.

Theorem 4.2 *If there exists a feasible solution to a system of linear equations, then there exists a basic feasible solution.*

Proof Consider the equations

$$\sum_{j=1}^{n}x_j\boldsymbol{a}_j=\boldsymbol{b}\qquad(4.1)$$
$$x_j\geq 0\quad (j=1,\ \ldots,n)$$

We will proceed by induction on k, the number of columns. The theorem is certainly true for $k = 1$. We will now assume that the theorem holds true for $k < n$ and will examine the case of $k = n$. Notice that the variables x_j are strictly positive; otherwise, it reduces to the case of $k = n - 1$. By assumption, there is a feasible solution. If the $\boldsymbol{a}_j$ corresponding to x_j are independent, then the solution is basic by definition, and the theorem is proved. If these $\boldsymbol{a}_j$ are not independent, then there exists a nontrivial linear combination of the $\boldsymbol{a}_j$ such that

$$\sum_{j=1}^{n} \lambda_j \boldsymbol{a}_j = \boldsymbol{0} \tag{4.2}$$

We can assume that some λ_j are positive. If not, we can multiply (4.2) by -1. Among those positive λ_j, one of them must give maximal value to the ratio $\frac{\lambda_j}{x_j}$. Without loss of generality, let that ratio be $\frac{\lambda_n}{x_n}$, and denote its value by θ. The value θ is non-negative because λ_n and x_n are positive. So, we have

$$\theta = \max_j \frac{\lambda_j}{x_j} = \frac{\lambda_n}{x_n} > 0$$

Multiplying (4.2) by $\frac{1}{\theta}$ and subtracting it from (4.1), we have

$$\frac{1}{\theta} \sum_{j=1}^{n} \left(\theta - \frac{\lambda_j}{x_j} \right) x_j \boldsymbol{a}_j = \boldsymbol{b} \tag{4.3}$$

Therefore, (4.3) gives a linear non-negative combination of $\boldsymbol{a}_j$ that equals $\boldsymbol{b}$. However, notice that the coefficient of $\boldsymbol{a}_n$ is zero because $\theta = \frac{\lambda_n}{x_n}$. As such, (4.3) reduces to the case of fewer than n vectors, and we are done by the inductive hypothesis. □

The vertices of a convex set that is defined by linear equations and inequalities correspond to basic feasible solutions.

Theorem 4.3 *The set of all feasible solutions to a linear program is a convex set.*

Proof Let $\boldsymbol{x}_1$ and $\boldsymbol{x}_2$ be two feasible solutions, i.e.,

$$\boldsymbol{A}\boldsymbol{x}_1 = \boldsymbol{b} \quad (\boldsymbol{x}_1 \geq \boldsymbol{0})$$
$$\boldsymbol{A}\boldsymbol{x}_2 = \boldsymbol{b} \quad (\boldsymbol{x}_2 \geq \boldsymbol{0})$$

Then for $1 \geq \lambda \geq 0$, we have that $\lambda \boldsymbol{x}_1 + (1 - \lambda)\boldsymbol{x}_2 \geq \boldsymbol{0}$ because each of the two terms is a product of non-negative numbers. Furthermore,

$$A[\lambda \boldsymbol{x}_1 + (1-\lambda)\boldsymbol{x}_2] = \lambda A\boldsymbol{x}_1 + (1-\lambda)A\boldsymbol{x}_2 = \lambda \boldsymbol{b} + (1-\lambda)\boldsymbol{b} = \boldsymbol{b}$$

Therefore, $\lambda \boldsymbol{x}_1 + (1-\lambda)\boldsymbol{x}_2$ is also a feasible solution. □

Theorem 4.4 *If all feasible solutions of a linear program are bounded, any feasible solution is a linear convex combination of basic feasible solutions.*

Proof Feasible solutions form a compact convex set, where basic feasible solutions correspond to extreme points of the convex set. Since any point of a convex set is a linear convex combination of its extreme points,[1] any feasible solution is a linear convex combination of basic feasible solutions. □

Theorem 4.5 *If there exists an optimum solution, then there exists a basic optimum solution.*

Proof By assumption, the optimum solution is finite. A linear function that is a concave function will attain its global minimum at an extreme point of the convex set. Note that an extreme point corresponds to a basic solution. □

4.3 Examples of Linear Programs

The following examples will be referenced in the next few sections. It is advised that the reader sit with the examples and try to understand what row reductions are being carried out before moving forward.

Example 1

$$\begin{aligned} \min \quad & z = x_1 + x_2 + 2x_3 + x_4 \\ \text{subject to} \quad & x_1 + 2x_3 - 2x_4 = 2 \\ & x_2 + x_3 + 4x_4 = 6 \\ & x_1, x_2, x_3, x_4 \geq 0 \end{aligned} \tag{4.4}$$

Rewriting the objective function as an equation, we have

$$\begin{aligned} -z + x_1 + x_2 + 2x_3 + x_4 &= 0 \quad (0) \\ x_1 + 2x_3 - 2x_4 &= 2 \quad (1) \\ x_2 + x_3 + 4x_4 &= 6 \quad (2) \end{aligned}$$

[1] A feasible solution $\boldsymbol{x}^*$ is called an optimum solution if $\boldsymbol{c}\boldsymbol{x}^* \leq \boldsymbol{c}\bar{\boldsymbol{x}}$ for all feasible solutions $\bar{\boldsymbol{x}}$ and $-\infty < \boldsymbol{c}\boldsymbol{x}^*$.

$$\begin{array}{lr}
-z \quad - x_3 - x_4 = -8 & (0) - (1) - (2) = (0') \\
x_1 \quad + 2x_3 - 2x_4 = 2 & (1) = (1') \\
x_2 + x_3 + 4x_4 = 6 & (2) = (2') \\
-z + \frac{1}{2}x_1 \quad - 2x_4 = -7 & (0') + \frac{1}{2}(1') = (0'') \\
\frac{1}{2}x_1 \quad + x_3 - x_4 = 1 & \frac{1}{2}(1') = (1'') \\
-\frac{1}{2}x_1 + x_2 \quad + 5x_4 = 5 & (2') - (1'') = (2'') \\
-z + \frac{1}{2}x_1 \quad - 2x_4 = -7 & (0'') = (0''') \\
\frac{1}{2}x_1 \quad + x_3 - x_4 = 1 & (1'') = (1''') \\
-\frac{1}{10}x_1 + \frac{1}{5}x_2 \quad + x_4 = 1 & \frac{1}{5}(2'') = (2''') \\
-z + \frac{3}{10}x_1 + \frac{2}{5}x_2 \quad = -5 & (0''') + 2(2''') = (0'''') \\
\frac{4}{10}x_1 + \frac{1}{5}x_2 + x_3 \quad = 2 & (1''') + (2''') = (1'''') \\
-\frac{1}{10}x_1 + \frac{1}{5}x_2 \quad + x_4 = 1 & (2''') = (2'''')
\end{array}$$

$x_1 = 0, \quad x_2 = 0, \quad x_3 = 2, \quad x_4 = 1,$ and $z = 5.$

Example 2

$$\begin{aligned}
\max \quad & x_1 + 2x_2 \\
\text{subject to} \quad & -x_1 + x_2 \le 6 \\
& x_2 \le 8 \\
& x_1 + x_2 \le 12 \\
& 4x_1 + x_2 \le 36 \\
& x_1, x_2 \ge 0
\end{aligned} \tag{4.5}$$

Rewriting the objective function as an equation and the constraints as equations with slack variables, we have

$$\begin{array}{lr}
x_0 - x_1 - 2x_2 \qquad = 0 & (1) \\
-x_1 + x_2 + s_1 \qquad = 6 & (2) \\
x_2 \quad + s_2 \qquad = 8 & (3) \\
x_1 + x_2 \quad + s_3 \quad = 12 & (4) \\
4x_1 + x_2 \quad + s_4 = 36 & (5)
\end{array}$$

$$
\begin{aligned}
&x_0 + 7x_1 \qquad\qquad\qquad + 2s_4 = 72 && (1) + 2(5) = (1')\\
&\quad - 5x_1 \quad + s_1 \qquad\qquad - s_4 = -30 && (2) - (5) = (2')\\
&\quad - 4x_1 \qquad + s_2 \qquad - s_4 = -28 && (3) - (5) = (3')\\
&\quad - 3x_1 \qquad\qquad + s_3 - s_4 = -24 && (4) - (5) = (4')\\
&\quad 4x_1 + x_2 \qquad\qquad + s_4 = 36 && (5) = (5')\\
&x_0 \qquad\qquad + \frac{7}{3}s_3 - \frac{1}{3}s_4 = 16 && (1') + \frac{7}{3}(4') = (1'')\\
&\qquad s_1 \qquad - \frac{5}{3}s_3 + \frac{2}{3}s_4 = 10 && (2') - \frac{5}{3}(4') = (2'')\\
&\qquad\qquad s_2 - \frac{4}{3}s_3 + \frac{1}{3}s_4 = 4 && (3') - \frac{4}{3}(4') = (3'')\\
&\quad x_1 \qquad\qquad - \frac{1}{3}s_3 + \frac{1}{3}s_4 = 8 && -\frac{1}{3}(4') = (4'')\\
&\qquad x_2 \qquad\quad + \frac{4}{3}s_3 - \frac{1}{3}s_4 = 4 && (5') + \frac{4}{3}(4') = (5'')\\
&x_0 \qquad\qquad + s_2 + s_3 \qquad = 20 && (1'') + (3'') = (1''')\\
&\qquad s_1 - 2s_2 + s_3 \qquad = 2 && (2'') - 2(3'') = (2''')\\
&\qquad\qquad 3s_2 - 4s_3 + s_4 = 12 && 3(3'') = (3''')\\
&\quad x_1 \qquad - s_2 + s_3 \qquad = 4 && (4'') - (3'') = (4''')\\
&\qquad x_2 \quad + s_2 \qquad\qquad = 8 && (5'') + (3'') = (5''')
\end{aligned}
$$

$x_1 = 4,\quad x_2 = 8,\quad s_1 = 2,\quad s_2 = 0,\quad s_3 = 0,\quad s_4 = 12,\quad$ and $\quad x_0 = 20.$

4.4 Pivot Operation

In the last section, you may have noticed that for both examples, the row reductions carried out at each step resulted in a change of basic variables. It turns out that at each step we applied the pivot operation.

In short, the *pivot operation* swaps a basic variable and a non-basic variable. So, in Example 2 from Section 4.3, notice how we started out with s_1, s_2, s_3, s_4 as our basic variables, and after the first step, we had basic variables x_2, s_1, s_2, s_3. This means that we applied the pivot operation in order to switch out s_4 from our basis with x_2.

Definition We call the column of the non-basic variable that we are swapping into our basis the *pivot column*, we call the row that contains the nonzero basic variable the *pivot row*, and we call the intersection of the pivot column and the pivot row the *pivot*.

In order to swap the basic variable with the non-basic variable, we scale the pivot row so that the coefficient of the non-basic variable is 1. Then, we add multiples of the pivot row to all other rows so that they have a coefficient of 0 in the pivot column. This is the pivot operation.

4.5 Simplex Method

We have studied the fundamental theorems of linear program in the previous chapters. Now, we shall develop computational methods.

Definition A linear program is in *standard form* if the constraints (not including $x_j \geq 0$) are all equations.

Definition A linear program is in *canonical form* if the constraints are all inequalities.

Theorem 4.2 states that if there is a feasible solution, then there is a basic feasible solution. Theorem 4.5 states that if there is an optimum solution, then there is a basic optimum solution. So, to get an optimum solution, we can choose m columns at a time from the matrix $\boldsymbol{A}$ and solve the m simultaneous equations for the m variables. However, this approach requires that $\binom{n}{m}$ sets of simultaneous equations be solved, which is not practical unless n is small. In general, the most practical method is the *Simplex Method*, which takes a very small number of steps to solve a linear program.

Let us assume that we have n variables and m constraint equations in a linear program, with $n > m$. Then, the solution space[2] is of $n - m$ dimensions. When we introduce a slack variable to convert an inequality into an equation, we add both one variable and one equation. As such, the dimension of the solution space remains the same. Thus, we have the freedom to assign arbitrary values to $(n - m)$ variables and can determine the values of the remaining m variables by solving m simultaneous equations in m variables. The variables with arbitrarily assigned values are non-basic variables. The variables with their values determined by solving simultaneous equations are basic variables. In general, there is an optimal basic solution with linearly independent vectors, where the non-basic variables are each assigned the value zero.

In a sense, the Simplex Method finds an optimum set of basic variables by iteratively exchanging non-basic and basic variables using the pivot operation. In Example 1 from Section 4.3, we first decide to choose x_1 and x_2 as non-basic variables and assign their values to be zero since the equations are in diagonal form

[2] Solution space is also called *solution set*.

with respect to x_1 and x_2. Then we successively pick different sets of variables to be non-basic variables until we find that the value of the objective function cannot be decreased.

Let us summarize the Simplex Method as follows. Assume all constraints are converted into m equations, and the n variables are all restricted to be non-negative $(m \le n)$.

1. We choose a set of m variables to be basic variables and use the Gauss-Jordan Method to put the constraints into diagonal form with respect to these m basic variables. If the first m variables are chosen to be basic variables, then the constraint equations are of the following form after applying the Gauss-Jordan Method. We use $\overline{b}_i$ to indicate the basic solution as compared with b_i in the linear program.

$$\begin{array}{llll} x_1 + & \cdots & + a_{1,m+1}x_{m+1} + \cdots + a_{1,n}x_n & = \overline{b}_1 \\ & x_2 & & \\ & \ddots & & \\ & & x_m + a_{m,m+1}x_{m+1} + \cdots + a_{m,n}x_n & = \overline{b}_m \end{array}$$

We tentatively assume that all $\overline{b}_i$ are non-negative so that we have a feasible basic solution.

$$\begin{aligned} x_i &= \overline{b}_i \quad (i = 1, \ldots, m) \\ x_j &= 0 \quad (j = m + 1, \ldots, n) \end{aligned}$$

2. Let the objective function be either

$$\max\ w = \sum c_j x_j$$

or

$$\min\ z = \sum c_j x_j$$

We shall eliminate the basic variables from the objective function so that the objective function is of the form, say

$$-z + \sum \overline{c}_j x_j = -60 \quad (j = m + 1, \ldots, n)$$

or

$$w + \sum \overline{c}_j x_j = 100 \quad (j = m + 1, \ldots, n)$$

In both cases, if all $\overline{c}_j \ge 0$, then increasing the value of non-basic variables will not improve the objective function, and $z = 60$ or $w = 100$ will not be the value of the objective function, which is actually the optimal value. If all $\overline{c}_j > 0$, then all x_j must be zero to obtain $z = 60$ or $w = 100$. Assuming that the condition $\overline{c}_j > 0$ implies optimality has become standard practice, and we use $(-z)$ and $(+w)$ in conformance with this practice.

3. If not all $\overline{c}_j \geq 0$ in the objective function, we choose the variable with the most negative coefficient, say $\overline{c}_s < 0$, and increase x_s from zero to the largest value possible. We use the ratio test to determine how much x_s can be increased without causing any current basic variables to become negative. The ratio test is

$$\min_i \left| \frac{\overline{b}_i}{a_{is}} \right| = \frac{\overline{b}_r}{a_{rs}}, \quad \forall a_{is} > 0$$

Then we apply the pivot operation with a_{rs} as our pivot. Recall that this changes x_s from non-basic to basic and changes x_r from basic to non-basic. Again, the coefficients of all basic variables are zero in the objective function, and we perform the optimality test.

Note that we can choose any $\overline{c}_s < 0$ and let x_s be the new basic variable. But once x_s is chosen, the new non-basic variable is determined by the ratio test.

When we solve simultaneous equations, we can work with the coefficients only, and we need not write down the names of the variables. This technique is used in linear programming: the computational framework is condensed into the so-called *Simplex tableau*, where the names of the variables are written along the outside edges. To illustrate the use of the tableau, we shall repeat Example 1 again.

$$\begin{aligned} \min \quad & z = x_1 + x_2 + 2x_3 + x_4 \\ \text{subject to} \quad & x_1 \quad\; + 2x_3 - 2x_4 = 2 \\ & x_2 + \; x_3 + 4x_4 = 6 \\ & x_1, x_2, x_3, x_4 \geq 0 \end{aligned} \tag{4.6}$$

We shall write the data into a Simplex tableau as

	$-z$	x_1	x_2	x_3	x_4	RHS
$-z$	1	1	1	2	1	0
x_1	0	1	0	2	−2	2
x_2	0	0	1	1	4	6

Note that in the above Simplex tableau, the objective function is written above the tableau (i.e., the 0th row); the current basic variables x_1 and x_2 are written to the left of the tableau.

If we subtract the first row and the second row from the 0th row expressing the objective function, we have

	$-z$	x_1	x_2	$\downarrow$ x_3	x_4	RHS
$-z$	1	0	0	-1	-1	-8
$\rightarrow x_1$	0	1	0	2	-2	2
x_2	0	0	1	1	4	6

	$-z$	x_1	x_2	x_3	$\downarrow$ x_4	RHS
$-z$	1	$\frac{1}{2}$	0	0	-2	-7
x_3	0	$\frac{1}{2}$	0	1	-1	1
$\rightarrow x_2$	0	$-\frac{1}{2}$	1	0	5	5

	$-z$	x_1	x_2	x_3	x_4	RHS	
$-z$	1	$\frac{3}{10}$	$\frac{2}{5}$	0	0	-5	$z = 5$
x_3	0	$\frac{4}{10}$	$\frac{1}{5}$	1	0	2	$x_3 = 2$
x_4	0	$-\frac{1}{10}$	$\frac{1}{5}$	0	1	1	$x_4 = 1$

Note that the labels along the upper boundary remain the same during the computation. Each row represents an equation; the equation is obtained by multiplying the corresponding coefficients by the variables. The double vertical lines separate the left-hand side (LHS) from the right-hand side (RHS). The labels along the left boundary of the tableau are the current basic variables; their values are equal to those on the right-hand side.

$$\begin{aligned} \max \quad & w = 3\pi_1 + 5\pi_2 \\ \text{subject to} \quad & 3\pi_1 + \ \pi_2 \leq 15 \\ & \pi_1 + \ \pi_2 \leq 7 \\ & \pi_2 \leq 4 \\ & -\pi_1 + 2\pi_2 \leq 6 \end{aligned} \tag{4.7}$$

	w	π_1	↓ π_2	s_1	s_2	s_3	s_4	RHS
w	1	−3	−5					0
s_1		3	1	1				15
s_2		1	1		1			7
s_3		0	1			1		4
→ s_4		−1	2				1	6

	w	↓ π_1	π_2	s_1	s_2	s_3	s_4	RHS
w	1	$-\frac{11}{2}$	0				$\frac{5}{2}$	15
s_1		$\frac{7}{2}$	0	1			$-\frac{1}{2}$	12
s_2		$\frac{3}{2}$	0		1		$-\frac{1}{2}$	4
→ s_3		$\frac{1}{2}$	0			1	$-\frac{1}{2}$	1
π_2		$-\frac{1}{2}$	1				$\frac{1}{2}$	3

	w	π_1	π_2	s_1	s_2	s_3	↓ s_4	RHS
w	1	0	0			11	−3	26
s_1		0	0	1		−7	3	5
→ s_2		0	0		1	−3	1	1
π_1		1	0			2	−1	2
π_2		0	1			1	0	4

	w	π_1	π_2	s_1	s_2	s_3	s_4	RHS
w	1	0	0		3	2	0	29
s_1		0	0	1	−3	2	0	2
s_4		0	0	0	1	−3	1	1
π_1		1	0		1	−1	0	3
π_2		0	1		0	1	0	4

4.6 Review and Simplex Tableau

In the computational aspect of linear algebra, the goal is to find the solution to simultaneous equations, a point of zero dimension. When the number of variables n is more than the number of equations m, the solution space has $n - m$ dimensions. When we start with any $n \geq 2$ variables and one constraint, we use the Ratio Method. The Ratio Method gives the optimum solution, which has one variable with a positive value, all other variables equal to zero. When there are $m \geq 2$

inequalities, the solution space is a polygon or a polytope—in general, a convex set. The optimum solution always occurs at one of the extreme points.

Since the objective function is a linear function, it is both a convex function and a concave function. Because it is a convex function, we can use the greedy approach, and because it is a concave function, we can stop at an extreme point.

The linear program is expressed as

$$\begin{aligned} \min \quad & z = \boldsymbol{cx} \\ \text{subject to} \quad & \boldsymbol{Ax} = \boldsymbol{b} \\ & \boldsymbol{x} \geq \boldsymbol{0} \end{aligned}$$

or

$$\begin{aligned} \max \quad & x_0 = \boldsymbol{cx} \\ \text{subject to} \quad & \boldsymbol{Ax} = \boldsymbol{b} \\ & \boldsymbol{x} \geq \boldsymbol{0} \end{aligned}$$

While the RHS $\boldsymbol{b}$ has m components, $m < n$, we need to find $x_1, x_2, \ldots, x_m$ positive variables and let the rest of the variables $x_{m+1}, \ldots, x_n$ be zero. In other words, the basis has m columns all associated with positive values, while variables associated with $(n - m)$ columns are zero.

In terms of the Homemaker Problem, the homemaker needs to buy m items if he needs m kinds of vitamins. For the Pill Salesperson Problem, the pill salesperson needs to fix the prices of m kinds of vitamin pills to compete with the supermarket.

In the Simplex tableau, there are $m + 1$ rows and $n + 2$ columns, with the rows associated with the current basic variables and the 0th row associated with $-z$ or x_0. The $(n + 2)$ columns are associated with all variables (including $-z$ or x_0), with the rightmost column representing vector $\boldsymbol{b}$.

To generalize the Ratio Method, we need to find the denominator of the ratio or multiply all columns of the matrix $\boldsymbol{A}$ by the inverse of a square matrix of size $(m + 1)$ by $(m + 1)$. This would mean that we need to choose an $m \times m$ square matrix. However, we can select the most promising column $\boldsymbol{a}_s$ (the column with the most negative coefficient) to enter the basis, to replace the rth row and make $a_{rs} = 1$. The operation needs a test to maintain that all components of RHS remain positive.

The beauty of the Simplex Method is that the process of replacing a square matrix by another matrix which differs in only one column is quite simple. As a folklore, we need to execute this process about $2m$ to $3m$ times, independent of n even if $n \gg m$.

Let us put the following linear program into the tableau form (Tableau 4.1).

$$\begin{aligned} \max \quad & x_0 \\ \text{subject to} \quad & x_0 \qquad\quad + 2x_3 - 2x_4 - x_5 = 0 \\ & \quad\; x_1 \qquad - 2x_3 + x_4 + x_5 = 4 \\ & \qquad\quad x_2 + 3x_3 - x_4 + 2x_5 = 2 \end{aligned}$$

Tableau 4.1

	x_0	x_1	x_2	x_3	x_4 ↓	x_5	RHS
x_0	1			2	−2	−1	0
→ x_1		1		−2	1*	1	4
x_2			1	3	−1	2	2

Tableau 4.2

	x_0	x_1	x_2	x_3 ↓	x_4	x_5	RHS
x_0	1	2	0	−2	0	1	8
x_4	0	1	0	−2	1	1	4
→ x_2	0	1	1	1*	0	3	6

Tableau 4.3

	x_0	x_1	x_2	x_3	x_4	x_5	RHS
x_0	1	4	2	0	0	7	20
x_4	0	3	2	0	1	7	16
x_3	0	1	1	1	0	3	6

There are two negative coefficients, -2 and -1, in the 0th row, so we select the column under x_4 to enter the basis, indicated by ↓. Because the fourth column has only one positive entry, we do not need to perform the feasibility test. The entry for pivot is indicated by an asterisk* on its upper right corner. Note that the labels on the left of the tableau are the current basic variables.

In Tableau 4.2, the only negative coefficient is -2 under x_3, and under this there is only one positive coefficient eligible for pivoting, indicated by an asterisk* to its upper right. The result is shown in Tableau 4.3.

In Tableau 4.3, all coefficients in the 0th row are non-negative. As such, we conclude that an optimum result is $x_0 = 20$, $x_4 = 16$, and $x_3 = 6$.

Note that $\boldsymbol{B} \cdot \boldsymbol{B}^{-1} = \boldsymbol{I}$, $\boldsymbol{B}^{-1}\boldsymbol{a}_j = \overline{\boldsymbol{a}}_j$, and $\boldsymbol{B}^{-1}\boldsymbol{b} = \overline{\boldsymbol{b}}$ where $\boldsymbol{B}$ is a corresponding basis matrix and $\boldsymbol{B}^{-1}$ is an inverse matrix of $\boldsymbol{B}$. We shall discuss this more in Chapter 6.

$$\begin{bmatrix} 1 & -2 & 0 \\ 0 & 1 & 0 \\ 0 & -1 & 1 \end{bmatrix} \begin{bmatrix} 1 & 2 & 0 \\ 0 & 1 & 0 \\ 0 & 1 & 1 \end{bmatrix} = \begin{bmatrix} 1 & 0 & 0 \\ 0 & 1 & 0 \\ 0 & 0 & 1 \end{bmatrix}$$

$$\begin{bmatrix} 1 & -2 & 2 \\ 0 & 1 & -2 \\ 0 & -1 & 3 \end{bmatrix} \begin{bmatrix} 1 & 4 & 2 \\ 0 & 3 & 2 \\ 0 & 1 & 1 \end{bmatrix} = \begin{bmatrix} 1 & 0 & 0 \\ 0 & 1 & 0 \\ 0 & 0 & 1 \end{bmatrix}$$

$$\begin{bmatrix} 1 & 2 & 0 \\ 0 & 1 & 0 \\ 0 & 1 & 1 \end{bmatrix} \begin{bmatrix} 2 \\ -2 \\ 3 \end{bmatrix} = \begin{bmatrix} -2 \\ -2 \\ 1 \end{bmatrix}$$

$$\begin{bmatrix} 1 & 4 & 2 \\ 0 & 3 & 2 \\ 0 & 1 & 1 \end{bmatrix} \begin{bmatrix} 0 \\ 4 \\ 2 \end{bmatrix} = \begin{bmatrix} 20 \\ 16 \\ 6 \end{bmatrix}$$

Now let us add another column to Tableau 4.1 and name it Tableau 4.4.

Here we choose the most negative coefficient and perform a feasibility test to choose the pivot. This gives us Tableaus 4.5. Tableaus 4.6 and 4.7 are similarly obtained.

This simple example shows that choosing the most negative coefficient may not be efficient. We use this example to show that we need the feasibility test $\left(\frac{4}{1}\right) > \left(\frac{2}{1}\right)$, and we need to choose the pivot as shown in Tableau 4.4. Up to now, there is no known "perfect" pivot selection rule.

In the Simplex Method, the choice of pivot is by the feasibility test. In rare cases, there is no improvement of the value of the basic feasible solution after the pivot. The way to cure it is called *Bland's rule*:

Tableau 4.4

	x_0	x_1	x_2	x_3	x_4	x_5	x_6	*RHS*
x_0	1			2	−2	−1	−3	0
x_1		1		−2	1	1	1	4
x_2			1	3	−1	2	1*	2

Tableau 4.5

	x_0	x_1	x_2	x_3	x_4	x_5	x_6	*RHS*
x_0	1	0	3	11	−5	5	0	6
x_1	0	1	−1	−5	2*	−1	0	2
x_6	0	0	1	3	−1	2	1	2

Tableau 4.6

	x_0	x_1	x_2	x_3	x_4	x_5	x_6	RHS
x_0	1	$\frac{5}{2}$	$\frac{1}{2}$	$-\frac{3}{2}$	0	$\frac{5}{2}$	0	11
x_4	0	$\frac{1}{2}$	$-\frac{1}{2}$	$-\frac{5}{2}$	1	$-\frac{1}{2}$	0	1
x_6	0	$\frac{1}{2}$	$\frac{1}{2}$	$\frac{1^*}{2}$	0	$\frac{3}{2}$	1	3

Tableau 4.7

	x_0	x_1	x_2	x_3	x_4	x_5	x_6	RHS
x_0	1	4	2	0	0	7	3	20
x_4	0	3	2	0	1	7	5	16
x_3	0	1	1	1	0	3	2	6

1. In the case of a tie in the feasibility test, select the column with the smallest index j to enter the basis.
2. If there are two or more columns which could be dropped out, select the column which has the smallest index j.

4.7 Linear Programs in Matrix Notation

We shall write a linear program in matrix notation as

$$\begin{aligned} \min \quad & z = \boldsymbol{cx} \\ \text{subject to} \quad & \boldsymbol{Ax} = \boldsymbol{b} \\ & \boldsymbol{x} \geq \boldsymbol{0} \end{aligned} \tag{4.8}$$

Assume that $\boldsymbol{A} = [\boldsymbol{B}, \boldsymbol{N}]$, where $\boldsymbol{B}$ is a feasible basis. Then the program (4.8) can be written as

$$\min \quad z = \boldsymbol{c}_B\boldsymbol{x}_B + \boldsymbol{c}_N\boldsymbol{x}_N \tag{4.9}$$

$$\begin{aligned} \text{subject to} \quad & \boldsymbol{Bx}_B + \boldsymbol{Nx}_N = \boldsymbol{b} \\ & \boldsymbol{x}_B, \boldsymbol{x}_N \geq \boldsymbol{0} \end{aligned} \tag{4.10}$$

If we solve for $\boldsymbol{x}_B$ in (4.10), we first discover that

$$\boldsymbol{Ix}_B = \boldsymbol{B}^{-1}\boldsymbol{b} - \boldsymbol{B}^{-1}\boldsymbol{Nx}_N$$

Substituting the value $\boldsymbol{x}_B$ into (4.9), we get

$$z = \boldsymbol{c}_B\boldsymbol{B}^{-1}\boldsymbol{b} + \left(\boldsymbol{c}_N - \boldsymbol{c}_B\boldsymbol{B}^{-1}\boldsymbol{N}\right)\boldsymbol{x}_N$$

Using $\boldsymbol{\pi} = (\pi_1, \pi_2, \ldots, \pi_m)$ to denote the expression $\boldsymbol{c}_B\boldsymbol{B}^{-1}$, we can simplify this to

$$z = \boldsymbol{\pi}\boldsymbol{b} + (\boldsymbol{c}_N - \boldsymbol{\pi}\boldsymbol{N})\boldsymbol{x}_N$$

If we let $\boldsymbol{x}_N = \boldsymbol{0}$, then $z = \boldsymbol{\pi}\boldsymbol{b}$. Explicitly, we see that at any stage of computation, the value of the objective function is equal to the current $\boldsymbol{\pi}$ multiplied by the original right-hand $\boldsymbol{b}$. This provides an independent check on the calculation.

4.8 Economic Interpretation of the Simplex Method

We can also write the linear program (4.8) in the form

$$\begin{aligned} \min \quad & z = \sum c_j x_j \\ \text{subject to} \quad & \sum a_{ij} x_j = b_i \quad (i = 1, \ldots, m) \\ & x_j \geq 0 \quad (j = 1, \ldots, n) \end{aligned} \tag{4.11}$$

Now, if we select the first m variables to be basic variables, and express the objective function z in terms of the non-basic variables,

$$z = z_0 + \sum \bar{c}_j x_j$$

where

$$\begin{aligned} \bar{c}_j &= 0 \quad (j = 1, \ldots, m) \\ \bar{c}_j &= c_j - \boldsymbol{\pi}\boldsymbol{a}_j \quad (j = m + 1, \ldots, n) \end{aligned}$$

then, when $\bar{c}_j \geq 0$ for all j, we have the optimum solution.

We can think of $\pi_i \; (i = 1, \ldots, m)$ as the price that we pay for producing one unit of the ith material. Let $\boldsymbol{a}_j$ denote any activity vector that is not in the basis. If we produce one unit of the vector $\boldsymbol{a}_j$, then a_{ij} units of the ith material will be produced. Since each unit of the ith material costs π_i dollars, the saving that could be achieved by using $\boldsymbol{a}_j$ is $\boldsymbol{\pi}\boldsymbol{a}_j$. The actual cost of operating $\boldsymbol{a}_j$ at unit level is c_j. Therefore if $c_j - \boldsymbol{\pi}\boldsymbol{a}_j \geq 0$, then that activity should not be used. If $c_j - \boldsymbol{\pi}\boldsymbol{a}_j < 0$, that means $\boldsymbol{a}_j$ would bring us some saving in the present system.

Definition We call $\bar{c}_j = c_j - \boldsymbol{\pi}\boldsymbol{a}_j$ the *relative cost* (or *reduced cost*).

Notice that if the relative cost is negative, or $\bar{c}_j = c_j - \boldsymbol{\pi}\boldsymbol{a}_j < 0$, then we should bring the corresponding x_j into the basis.

Definition This operation of changing the basis is the result of what is sometimes called the *pricing operation*. In the economic interpretation it corresponds to lowering the price of an underutilized resource.

Definition π_i is called the *shadow price* of each row under the present basis.

Notice that a basis $\boldsymbol{B}$ is an optimum basis if $\boldsymbol{B}^{-1}\boldsymbol{b} \geq \boldsymbol{0}$ and $c_j - \boldsymbol{c}_B\boldsymbol{B}^{-1}\boldsymbol{a}_j \geq 0$ for all j. Also, notice that π_i gives the rate of change of z as b_i varies.

4.9 Summary of the Simplex Method

The Simplex Method can be described in the following steps:

1. Start with a tableau where $a_{i0} \geq 0 \ \ (i = 1, \ldots, m)$.
2. If all $a_{0j} \geq 0$, stop. The current solution is the optimum solution. Otherwise, among $a_{0j} < 0$, let $a_{0s} = \min_j a_{0j} < 0$, i.e., a_{0s} is the most negative coefficient. This column s is our pivot column.
3. Among $a_{is} > 0$, let $a_{r0}/a_{rs} = \min_i a_{i0}/a_{is} \ \ (i = 1, \ldots, m)$, i.e., among the positive coefficients in the column s, we take all the ratios. The row r is our pivot row, and a_{rs} is our pivot.
4. Perform the pivot operation with a_{rs} as the pivot. Change the basic variable x_r into x_s on the left of the tableau. Return to Step 2.

In Step 2, we could choose any $a_{0j} = \overline{c}_j < 0$ and the algorithm will still work. The criterion $\min_j \overline{c}_j < 0$ is used analogous to the steepest descent (greedy) algorithm. In Step 3, the test $\min_i a_{i0}/a_{is}$ for selecting the pivot row is called the *ratio test*. It is used to maintain $a_{i0} \geq 0$.

The three steps are repeated until in Step 2 there are no negative coefficients in the 0th row. Then the optimum solution is obtained by setting the basic variables x_i on the left of the tableau equal to a_{i0} in that current tableau.

4.10 Exercises

1. An equation can be replaced by two inequalities. How many inequalities can replace two equations? What is a slack variable?
2. What does it mean when a system of simultaneous equations is consistent? When a system of simultaneous equations is redundant? When a solution is degenerate? Is the following system of equations redundant?

$$\begin{aligned} x_1 + \ x_2 + \ x_3 &= 30 \\ 2x_1 + 2x_2 + 2x_3 &= 60 \\ x_1 + 2x_2 + 3x_3 &= 60 \end{aligned}$$

3. Use slack variables to change the following inequalities into a system of equations:

$$\begin{aligned} \max \quad & 60x + 40y \\ \text{subject to} \quad & 5x + 2y \leq 52 \\ & \ x + \ y \leq 20 \\ & x, y \geq 0 \end{aligned}$$

4. Change the following equations into a system of inequalities:

$$\begin{aligned} 12x + 5y + u &= 30 \\ x + 2y + v &= 12 \\ -3x - 2y + w &= 0 \end{aligned}$$

5. State whether the following sets of vectors are linearly dependent or independent:

(a) $\begin{bmatrix} 4 \\ 2 \end{bmatrix}, \begin{bmatrix} 5 \\ 3 \end{bmatrix}$

(b) $\begin{bmatrix} 1 \\ -3 \\ 2 \end{bmatrix}, \begin{bmatrix} -2 \\ 1 \\ 1 \end{bmatrix}, \begin{bmatrix} 1 \\ 7 \\ -8 \end{bmatrix}$

(c) $\begin{bmatrix} 2 \\ 3 \\ 4 \end{bmatrix}, \begin{bmatrix} 7 \\ 5 \\ 4 \end{bmatrix}, \begin{bmatrix} 1 \\ 0 \\ 9 \end{bmatrix}, \begin{bmatrix} 5 \\ 5 \\ 8 \end{bmatrix}$

(d) $\begin{bmatrix} 5 \\ 4 \\ 1 \end{bmatrix}, \begin{bmatrix} 1 \\ 2 \\ 3 \end{bmatrix}, \begin{bmatrix} -2 \\ 1 \\ 8 \end{bmatrix}$

6. Consider the following linear program:

$$\begin{aligned} \max \quad & x_0 = x_1 + x_2 + x_3 + x_4 \\ \text{subject to} \quad & \begin{bmatrix} 9 \\ 1 \end{bmatrix} x_1 + \begin{bmatrix} 1 \\ 9 \end{bmatrix} x_2 + \begin{bmatrix} 5 \\ 4 \end{bmatrix} x_3 + \begin{bmatrix} 4 \\ 8 \end{bmatrix} x_4 = \begin{bmatrix} 11 \\ 11 \end{bmatrix} \\ & x_j \geq 0 \quad (j = 1, 2, 3, 4) \end{aligned}$$

Without doing the actual computations, could you find the optimal basic variables? Use a graph if you need it.

7. Solve the following linear programming problem and draw the solution space graphically. How would you describe this particular solution space?

$$\begin{aligned} \max \quad & 8x + 5y \\ \text{subject to} \quad & x - y \leq 350 \\ & x \geq 200 \\ & y \geq 200 \\ & x, y \geq 0 \end{aligned}$$

8. For the following matrices, apply the pivot operation to the bolded elements:

(a) $\begin{bmatrix} -2 & 3 & \mathbf{3} \\ -2 & 4 & 2 \\ 3 & 5 & 1 \end{bmatrix}$

(b) $\begin{bmatrix} 1 & 2 & 2 & 8 \\ 0 & \mathbf{3} & -5 & 1 \\ 6 & 5 & 3 & 5 \end{bmatrix}$

9. Do the labels (names) of all variables above the Simplex tableau change during the computation? Do the labels (names) of variables to the left of the Simplex tableau change? Are these names of the current basic variables or of the current non-basic variables?

10. Solve the following linear programming problems using the Simplex Method:

(a)
$$\begin{aligned} \max \quad & 2x + y \\ \text{subject to} \quad & 4x + y \leq 150 \\ & 2x - 3y \leq -40 \\ & x, y \geq 0 \end{aligned}$$

(b)
$$\begin{aligned} \min \quad & -6x - 4y + 2z \\ \text{subject to} \quad & x + y + 4z \leq 20 \\ & -5y + 5z \leq 100 \\ & x + 3y + z \leq 400 \\ & x, y, z \geq 0 \end{aligned}$$

(c)
$$\begin{aligned} \max \quad & 3x + 5y + 2z \\ \text{subject to} \quad & 5x + y + 4z \leq 50 \\ & x - y + z \leq 150 \\ & 2x + y + 2z \leq 100 \\ & x, y, z \geq 0 \end{aligned}$$

11. Solve the following linear programming problems using the Simplex Method. Note that some variables are unrestricted:

(a)
$$\begin{aligned} \min \quad & -x + y \\ \text{subject to} \quad & x + 2y \leq 100 \\ & 3x - 6y \leq 650 \\ & y \geq 0, \quad x \text{ unrestricted} \end{aligned}$$

(b)
$$\begin{aligned} \max \quad & x + y + 2z \\ \text{subject to} \quad & 5x + y + \ z \leq 240 \\ & x - y + \ z \leq -50 \\ & 2x + y + 2z \leq 400 \\ & x, z \geq 0, \quad y \text{ unrestricted} \end{aligned}$$

12. You need to buy pills in order to help with your vitamin and supplement deficiency in calcium, iron, and vitamin A. Each pill of Type I contains 6 units of calcium, 2 units of iron, and 3 units of vitamin A and costs \$0.10. Each pill of Type II contains 1 unit of calcium, 4 units of iron, and 7 units of vitamin A and costs \$0.12. You need a minimum of 250 units of calcium, 350 units of iron, and 420 units of vitamin A. How should you purchase pills of Type I and Type II in order to minimize your spending?
13. The K Company produces three types (A, B, C) of accessories which yield profits of \$7, \$5, and \$3, respectively. To manufacture an A accessory, it takes 3 minutes on machine I, 2 minutes on machine II, and 2 minutes on machine III. In order to manufacture a B accessory, it takes 1 minute, 3 minutes, and 1 minute on machines I, II, and III, respectively. In order to manufacture a C accessory, it takes 1 minute, 2 minutes, and 2 minutes on machines I, II, and III, respectively. There are 10 hours available on machine I, 7 hours on machine II, and 8 hours on machine III for manufacturing these accessories each day. How many accessories of each type should the K Company manufacture each day in order to maximize its profit?

Duality and Complementary Slackness 5

In this chapter, we will develop the concept of duality, as well as the related theorem of complementary slackness which not only tells us when we have optimal solutions, but also leads us to the Dual Simplex Method.

5.1 Primal and Dual Programs

In matrix notation, the minimization and maximization linear programs can be represented as

$$\begin{array}{llll} \min & \boldsymbol{cx} & \max & \boldsymbol{yb} \\ \text{subject to} & \boldsymbol{Ax} \geq \boldsymbol{b} & \text{subject to} & \boldsymbol{yA} \leq \boldsymbol{c} \\ & \boldsymbol{x} \geq \boldsymbol{0} & & \boldsymbol{y} \geq \boldsymbol{0} \end{array}$$

where $\boldsymbol{x}$ is a column vector with n components, $\boldsymbol{y}$ is a row vector with m components, $\boldsymbol{A}$ is an $m \times n$ matrix, $\boldsymbol{c}$ is a row vector with n components, and $\boldsymbol{b}$ is a column vector with m components.

Definition It turns out that for every linear program, there is another linear program closely associated with it. The two linear programs form a pair. If we call the first linear program the *primal program*, then we call the second linear program the *dual program*.

Most books denote the variables by a column vector $\boldsymbol{x} = [x_1, x_2, \ldots, x_n]$. In the dual program, the variables are either denoted by $\boldsymbol{\pi} = (\pi_1, \pi_2, \ldots, \pi_m)$, called the shadow prices (see Section 4.8), or denoted by a row vector $\boldsymbol{y} = (y_1, y_2, \ldots, y_m)$.

Definition We call the components of the row vector, $\boldsymbol{y} = (y_1, y_2, \ldots, y_m)$, the *dual variables*.

T.C. Hu, A.B. Kahng, *Linear and Integer Programming Made Easy*,
DOI 10.1007/978-3-319-24001-5_5

Let us now consider a linear program and its dual.

$$\begin{array}{llll} \min & z = x_1 + 2x_2 + 3x_3 & \max & w = 7y_1 + 11y_2 \\ \text{subject to} & 4x_1 + 5x_2 + 6x_3 \geq 7 & \text{subject to} & 4y_1 + 8y_2 \leq 1 \\ & 8x_1 + 9x_2 + 10x_3 \geq 11 & & 5y_1 + 9y_2 \leq 2 \\ & x_1, x_2, x_3 \geq 0 & & 6y_1 + 10y_2 \leq 3 \\ & & & y_1, y_2 \geq 0 \end{array}$$

A pair of feasible solutions to these programs is

$$x_1 = x_2 = x_3 = 1 \quad \text{with} \quad z = 6$$

and

$$y_1 = y_2 = \frac{1}{12} \quad \text{with} \quad w = \frac{3}{2}$$

Notice that the value of the minimized objective function z is greater than or equal to the value of the maximized objective function w. It turns out that this is always the case. This is analogous to the Max-Flow Min-Cut Theorem, which states that the value of a network cut is always greater than or equal to the value of the flow across that cut. In our example, the optimal solutions are

$$x_1 = \frac{7}{4}, \quad x_2 = x_3 = 0 \quad \text{with} \quad z = \frac{7}{4}$$

and

$$y_1 = \frac{1}{4}, \quad y_2 = 0 \quad \text{with} \quad w = \frac{7}{4}$$

The key thing to note in this example is that $z = w$.

In a linear program, such as the Homemaker Problem, we always write the objective function as an equation

$$\min z, \qquad -z + c_1x_1 + c_2x_2 + \cdots + c_nx_n = 0$$

In a maximization problem, the objective function is expressed as

$$\max x_0, \quad x_0 - c_1x_1 - c_2x_2 - \cdots - c_nx_n = 0$$

In both cases, the coefficients c_j can be either negative or positive. Notice that if we have all $c_j \geq 0$ in the tableau in the minimization problem, then the current value of z is optimum.

5.2 Duality and Complementary Slackness

Recall the Homemaker Problem and the Pill Salesperson Problem of Chapter 2. The homemaker and the pill salesperson wanted to minimize the cost and maximize the profit, respectively. These were not random examples. It turns out that most linear programs exist in pairs. Furthermore, many interesting relations can exist between a given pair of linear programs. Chief among these is the duality relationship. Consider the following pair of linear programs, shown side by side.

Canonical Form of Duality

Primal Program		Dual Program	
min	$z = \boldsymbol{cx}$	max	$w = \boldsymbol{yb}$
subject to	$\boldsymbol{Ax} \geq \boldsymbol{b}$	subject to	$\boldsymbol{yA} \leq \boldsymbol{c}$
	$\boldsymbol{x} \geq \boldsymbol{0}$		$\boldsymbol{y} \geq \boldsymbol{0}$

Standard Form of Duality

Primal Program		Dual Program	
min	$z = \boldsymbol{cx}$	max	$w = \boldsymbol{\pi b}$
subject to	$\boldsymbol{Ax} = \boldsymbol{b}$	subject to	$\boldsymbol{\pi A} \leq \boldsymbol{c}$
	$\boldsymbol{x} \geq \boldsymbol{0}$		$\boldsymbol{\pi} \lesseqgtr \boldsymbol{0}$

For ease of exposition, we shall use the canonical form in our discussion. Notice the symmetry between the two programs in the canonical form.

1. Each inequality of one program corresponds to a non-negative variable of the other program.
2. Each equation of the standard form of the duality corresponds to an unrestricted variable $\boldsymbol{\pi}$ of the dual program.

Theorem 5.1 (Theorem of Duality) *Given a pair of primal and dual programs (in canonical form), exactly one of the following cases must be true.*

1. Both programs have optimum solutions and their values are the same, i.e., $\min z = \max w \Leftrightarrow \min \boldsymbol{cx} = \max \boldsymbol{yb}$.
2. One program has no feasible solution, and the other program has at least one feasible solution, but no (finite) optimum solution.
3. Neither of the two programs has a feasible solution.

Cases 2 and 3 are easy to understand. To prove case 1, we apply the theorem of separating hyperplanes from Section 3.4.

Lemma 5.1 *If* $\overline{\boldsymbol{x}}$ *and* $\overline{\boldsymbol{y}}$ *are feasible solutions to a pair of primal and dual programs (shown as follows), then* $\boldsymbol{c}\overline{\boldsymbol{x}} \geq \overline{\boldsymbol{y}}\boldsymbol{b}$.

$$\begin{array}{llll} \min & \boldsymbol{cx} & \max & \boldsymbol{yb} \\ \text{subject to} & \boldsymbol{Ax} \geq \boldsymbol{b} & \text{subject to} & \boldsymbol{yA} \leq \boldsymbol{c} \\ & \boldsymbol{x} \geq \boldsymbol{0} & & \boldsymbol{y} \geq \boldsymbol{0} \end{array}$$

Proof Since $\overline{\boldsymbol{x}}$ is feasible, $\boldsymbol{A}\overline{\boldsymbol{x}} \geq \boldsymbol{b}$. Multiplying by $\overline{\boldsymbol{y}} \geq \boldsymbol{0}$ on both sides, we have $\overline{\boldsymbol{y}}\boldsymbol{A}\overline{\boldsymbol{x}} \geq \overline{\boldsymbol{y}}\boldsymbol{b}$. Since $\overline{\boldsymbol{y}}$ is feasible, $\overline{\boldsymbol{y}}\boldsymbol{A} \leq \boldsymbol{c}$. Multiplying by $\overline{\boldsymbol{x}} \geq \boldsymbol{0}$ on both sides, we have $\overline{\boldsymbol{y}}\boldsymbol{A}\overline{\boldsymbol{x}} \leq \boldsymbol{c}\overline{\boldsymbol{x}}$. Therefore, $\boldsymbol{c}\overline{\boldsymbol{x}} \geq \overline{\boldsymbol{y}}\boldsymbol{A}\overline{\boldsymbol{x}} \geq \overline{\boldsymbol{y}}\boldsymbol{b}$. □

Theorem 5.2 (Theorem of Weak Complementary Slackness) *Given a pair of primal and dual problems in canonical form, the necessary and sufficient conditions for a pair of feasible solutions* $\overline{\boldsymbol{x}}$ and $\overline{\boldsymbol{y}}$ *to be optimum solutions are*

$$\overline{\boldsymbol{y}}(\boldsymbol{A}\overline{\boldsymbol{x}} - \boldsymbol{b}) = 0$$
$$(\boldsymbol{c} - \overline{\boldsymbol{y}}\boldsymbol{A})\overline{\boldsymbol{x}} = 0$$

Proof Since $\overline{\boldsymbol{x}}$ and $\overline{\boldsymbol{y}}$ are both feasible, we have

$$\alpha = \overline{\boldsymbol{y}}(\boldsymbol{A}\overline{\boldsymbol{x}} - \boldsymbol{b}) \geq 0$$
$$\beta = (\boldsymbol{c} - \overline{\boldsymbol{y}}\boldsymbol{A})\overline{\boldsymbol{x}} \geq 0$$

because both α and β are products of non-negative terms. Furthermore, $\alpha + \beta = -\overline{\boldsymbol{y}}\boldsymbol{b} + \boldsymbol{c}\overline{\boldsymbol{x}} \geq 0$. From the theorem of duality, in order for $\overline{\boldsymbol{x}}$ and $\overline{\boldsymbol{y}}$ to be optimum, we must have $-\overline{\boldsymbol{y}}\boldsymbol{b} + \boldsymbol{c}\overline{\boldsymbol{x}} = 0$. This implies that $\alpha + \beta = 0$, and since $\alpha \geq 0$ and $\beta \geq 0$, this implies that $\alpha = 0$ and $\beta = 0$. □

Corollary of Theorem 5.2 *Given a pair of primal and dual problems in canonical form, the necessary and sufficient conditions for feasible solutions* $\overline{\boldsymbol{x}}$ and $\overline{\boldsymbol{y}}$ *to be optimum are that they satisfy the relations* (5.1)–(5.4).

We rewrite $\overline{\boldsymbol{y}}(\boldsymbol{A}\overline{\boldsymbol{x}} - \boldsymbol{b}) = 0$ and $(\boldsymbol{c} - \overline{\boldsymbol{y}}\boldsymbol{A})\overline{\boldsymbol{x}} = 0$ as

$$\overline{y}_i(\boldsymbol{a}_i\overline{\boldsymbol{x}} - b_i) = 0 \quad \text{for each } i$$
$$\left(c_j - \overline{\boldsymbol{y}}\boldsymbol{a}_j\right)\overline{x}_j = 0 \quad \text{for each } j$$

Since $\boldsymbol{a}_i\overline{\boldsymbol{x}} - b_i \geq 0$, in order for $\overline{y}_i(\boldsymbol{a}_i\overline{\boldsymbol{x}} - b_i) = 0$, at least one factor must be zero, i.e.,

$$\boldsymbol{a}_i\overline{\boldsymbol{x}} - b_i > 0 \quad \text{implies} \quad \overline{y}_i = 0 \tag{5.1}$$

$$\overline{y}_i > 0 \quad \text{implies} \quad \boldsymbol{a}_i\overline{\boldsymbol{x}} - b_i = 0 \tag{5.2}$$

Likewise

$$c_j - \overline{\boldsymbol{y}}\boldsymbol{a}_j > 0 \quad \text{implies} \quad \overline{x}_j = 0 \tag{5.3}$$

$$\overline{x}_j > 0 \quad \text{implies} \quad c_j - \overline{\boldsymbol{y}}\boldsymbol{a}_j = 0 \tag{5.4}$$

The relations (5.1)–(5.4) must hold true for every pair of optimum solutions. It may happen that both $\overline{y}_i = 0$ and $\boldsymbol{a}_i\overline{\boldsymbol{x}} - b_i = 0$ are true. The following theorem emphasizes that there will exist at least one pair of optimum solutions for which $\overline{y}_i = 0$ and $\boldsymbol{a}_i\overline{\boldsymbol{x}} - b_i = 0$ cannot happen at the same time.

Theorem 5.3 (Theorem of Strong Complementary Slackness) *Given a pair of primal and dual problems in canonical form both with feasible solutions, there exists at least one pair of optimum solutions* $\overline{\boldsymbol{x}}$ and $\overline{\boldsymbol{y}}$ *satisfying*

$$(A\overline{\boldsymbol{x}} - \boldsymbol{b}) + \overline{\boldsymbol{y}}^{\mathrm{T}} > 0$$
$$(\boldsymbol{c} - \overline{\boldsymbol{y}}A) + \overline{\boldsymbol{x}}^{\mathrm{T}} = 0$$

In more detail, we have

$$\boldsymbol{a}_i\overline{\boldsymbol{x}} - b_i = 0 \quad \text{implies} \quad \overline{y}_i > 0 \tag{5.5}$$

$$\overline{y}_i = 0 \quad \text{implies} \quad \boldsymbol{a}_i\overline{\boldsymbol{x}} - b_i > 0 \tag{5.6}$$

And also

$$c_j - \overline{\boldsymbol{y}}\boldsymbol{a}_j = 0 \quad \text{implies} \quad \overline{x}_j > 0 \tag{5.7}$$

$$\overline{x}_j = 0 \quad \text{implies} \quad c_j - \overline{\boldsymbol{y}}\boldsymbol{a}_j > 0 \tag{5.8}$$

These must be true for at least one pair of optimum solutions.

5.3 Economic and Geometric Interpretation

Economic Interpretation. For our economic interpretation, we can regard b_i as resources, and say that if a resource is not fully used (i.e., $a_{ij}x_j - b_i > 0$), then its price y_i is zero. Notice that if the price of a resource is positive, then the resource is all used up.

If x_j is the level of activity associated with the column $\boldsymbol{a}_j$, and if the cost of the activity c_j is too expensive (i.e., $c_j - y_j a_{ij} > 0$), then the level of the jth activity is zero $(x_j = 0)$. Furthermore, if the level of activity x_j is positive, then its cost must equal its worth at optimum, i.e., $c_j = y_i a_{ij}$.

In the Simplex Method, we first express z as a sum of $c_j x_j (j = 1, \ldots, n)$, and then we express z as $\sum \overline{c}_j x_j + k$ where k is a constant and $\overline{c}_j = c_j - a_{ij} y_i$. If $\overline{c}_j \geq 0$

for all j, then no improvements are possible. If $\overline{c}_j < 0$, then we should increase the level of that activity or make the non-basic variable basic.

In the Simplex Method, we successively interchange basic and non-basic variables until all non-basic variables are priced out, i.e., $\overline{c}_j \geq 0$ for all non-basic variables.

We can now more formally state the definition of the pricing operation from Section 4.8.

Definition The process of checking if a column $\boldsymbol{a}_j$ should be used by

$$\overline{c}_j = c_j - y_i a_{ij} \lesseqgtr 0$$

is called the *pricing operation*.

Geometric Interpretation. For our geometric interpretation, we can consider an inequality

$$a_{ij} x_j \geq b_i$$

as a half space. The intersection of all these half spaces defines the convex set which is the solution space (high dimension room).

For any point inside the solution space, we can always move the point along the gradient (steepest descent) to improve the value until the point is at a corner where the gradient is the convex combination of the normal vectors to the hyperplanes which define the corner point (extreme point). Notice that due to inequality constraints, we may have to move along an edge which minimizes the objective function but not at the quickest rate. Finally we may be at the intersection of two inequalities or the extreme point of the convex set. At that point, any local movement will increase the objective function.

5.4 Dual Simplex Method

Before we discuss the Dual Simplex Method, let us first review the Simplex Method. The two methods are tied together by duality theory. Consider a linear program in standard form

$$\begin{aligned} \min \quad & z = c_j x_j \quad (j = 1, \ldots, n) \\ \text{subject to} \quad & a_{ij} x_j = b_i \quad (i = 1, \ldots, m)(m \leq n) \\ & x_j \geq 0 \end{aligned}$$

Assume that we have put this into diagonal form with respect to $x_1, \ldots, x_m$, and $-z$. This gives

$$
\begin{aligned}
-z \ + \cdots \quad & + \bar{c}_{m+1}x_{m+1} + \cdots + \bar{c}_n x_n = -\bar{z}_0 \\
x_1 + \cdots \quad & + \bar{a}_{1,m+1}x_{m+1} + \cdots + \bar{a}_{1,n}x_n = \bar{b}_1 \\
x_2 \quad & \\
\ddots \quad & \\
x_m & + \bar{a}_{m,m+1}x_{m+1} + \cdots + \bar{a}_{m,n}x_n = \bar{b}_m
\end{aligned}
$$

The basic solution is $z = \bar{z}_0$, $x_i = \bar{b}_i \ (i = 1, \ldots, m)$, $x_{m+1} = x_{m+2} = \cdots = x_n = 0$. We shall assume that all $\bar{b}_i$ are non-negative so that we have a starting feasible basis.

Let us rewrite the diagonal form as (omitting the bar over coefficients)

$$
\begin{aligned}
x_0 + \cdots \quad & + a_{0,m+1}x_{m+1} + \cdots + a_{0,n}x_n = a_{00} \\
x_1 + \cdots \quad & + a_{1,m+1}x_{m+1} + \cdots + a_{1,n}x_n = a_{10} \\
x_2 \quad & \\
\ddots \quad & \\
x_m & + a_{m,m+1}x_{m+1} + \cdots + a_{m,n}x_n = a_{m0}
\end{aligned}
$$

where $x_0 = -z$, $a_{00} = -\bar{z}_0$, $a_{0j} = \bar{c}_j$, and $a_{i0} = \bar{b}_i$.

We then put this into the following tableau.

1	x_1	x_2	...	x_m	x_{m+1}	x_{m+2}	...	x_n
a_{00}	0	0	...	0	$a_{0,m+1}$	$a_{0,m+2}$	...	$a_{0,n}$
a_{10}	1	0	...	0	$a_{1,m+1}$	$a_{1,m+2}$	...	$a_{1,n}$
a_{20}	0	1	...	0	$a_{2,m+1}$	$a_{2,m+2}$	...	$a_{2,n}$
$\vdots$	$\vdots$	$\vdots$	$\ddots$	$\vdots$	$\vdots$	$\vdots$	$\ddots$	$\vdots$
a_{m0}	0	0	...	1	$a_{m,m+1}$	$a_{m,m+2}$	...	$a_{m,n}$

The top row of the tableau expresses x_0 in terms of all variables. Every row in the tableau is an equation. To the left of the tableau are the current basic variables. We start with a tableau in which $a_{i0} \geq 0 \ (i = 1, \ldots, m)$. This condition is called primal feasible.

Definition A *primal feasible* condition is a condition that implies that we have a feasible solution for the program which we want to solve. For example, the condition that the basic variables $= a_{i0}$ is primal feasible.

Definition The condition $a_{0j} \geq 0 \;\; (j = 1, \ldots, n)$ is called *dual feasible* because it implies that we have a feasible solution for the dual program.

Notice that if both $a_{i0} \geq 0 \;\; (i = 1, \ldots, m)$ and $a_{0j} \geq 0 \;\; (j = 1, \ldots, n)$, then the solution is optimum. To apply the Simplex Method, we start with the following tableau.

$-z$

w	x_1	x_2	x_m					x_n	
1	0	0	0	$\pm$	$\pm$	$\pm$	$\pm$	$\pm$	0
	1								$\oplus$
		1							$\oplus$
			1						$\oplus$

(Note that $\pm$ denotes unrestricted, and $\oplus$ denotes non-negative.)

Notice that the coefficients in the 0th row associated with the starting basic variables are zero, and we can read the values of the basic variables as the current $b_i \geq 0$ because the first m columns form an identity matrix $\boldsymbol{I}$. If all coefficients in the 0th equation are non-negative, then we have the optimum solution. If not, we first pick $\overline{c}_j < 0$, and use one ratio test to decide which basic variable to leave the basis, say $\overline{c}_s < 0$,

$$\frac{b_r}{a_{rs}} = \min \left| \frac{b_i}{a_{ij}} \right| \qquad \forall \; a_{ij} > 0$$

Then, we perform the pivot operation which makes $a_{rs} = 1$. The process is iterated until all $\overline{c}_j \geq 0$. If all $\overline{c}_j \geq 0$, the current solution is the optimum solution. Note that at every step of the Simplex Method, we keep a primal feasible solution, i.e., $\overline{b}_i \geq 0$, and at the end, $\overline{c}_j \geq 0$. The current solution is both primal and dual feasible, hence optimum.

The Dual Simplex Method is carried out in the same tableau as the primal Simplex Method. The Dual Simplex Method starts with a dual feasible solution and maintains its dual feasibility. It first decides what variable is to leave the basis and then decides which variable is to enter the basis. The Dual Simplex Method can be described in the following steps:

1. Start with a tableau in which $a_{0j} \geq 0$ for $j = 1, \ldots, n$.
2. If $\overline{b}_i = a_{0j} \geq 0$ for all $i = 1, \ldots, m$, then the program is solved. If not, select a $\overline{b}_r < 0$, and the variable x_r becomes a non-basic variable. Let

$$\max_j \frac{\overline{c}_j}{a_{rj}} = \frac{\overline{c}_s}{a_{rs}} \quad \left(a_{rj} < 0\right)$$
$$\text{or} \quad \min_j \left|\frac{\overline{c}_j}{a_{rj}}\right| = \left|\frac{\overline{c}_s}{a_{rs}}\right| \quad \left(a_{rj} < 0\right)$$

(We want $\overline{c}_j = c_j - \boldsymbol{\pi a}_j \geq 0$ to maintain the dual feasibility.)

3. Perform the pivot operation with a_{rs} as the pivot. Note that only negative elements are possible pivot elements.
4. Steps 2 and 3 are repeated until there is no b_i less than zero.

In the Dual Simplex Method, we start with the following tableau.

$-z$

w	x_1	x_2	x_m				x_n	
1	0	0	0	$\oplus$	$\oplus$	$\oplus$	$\oplus$	0
	1							$\pm$
		1						$\pm$
			1					$\pm$

(Note that $\pm$ denotes unrestricted, and $\oplus$ denotes non-negative.)

Since $b_i \lesseqgtr 0$, we select that row, and select x_r to leave the basis. To select the non-basic variable to enter the basis, we again perform a test like ratio testing and find

$$\min_j \left|\frac{\overline{c}_j}{a_{rj}}\right| = \left|\frac{\overline{c}_s}{a_{rs}}\right| \quad a_{rj} < 0$$

Again, the a_{rs} is made 1 by the pivot operation. The process is iterated until all $b_i \geq 0$.

We shall do the following example by the Dual Simplex Method:

$$\begin{aligned} \min \quad & z = -1 + 3x_2 + 5x_4 \\ \text{subject to} \quad & x_1 - 3x_2 \qquad - \; x_4 = -4 \\ & \qquad x_2 + x_3 + \; x_4 = 3 \\ & x_1, x_2, x_3, x_4 \geq 0 \end{aligned} \tag{5.9}$$

The system is already in diagonal form with respect to x_1 and x_3. Putting this in the usual Simplex tableau, we have Tableau 5.1. Note that $a_{0j} \geq 0 \;\, (j = 1, 2, 3, 4)$,

Tableau 5.1

	1	x_1	x_2 ↓	x_3	x_4
$-z$	1	0	3	0	5
→ x_1	−4	1	−3*	0	−1
x_3	3	0	1	1	1

and hence it is dual feasible. In the leftmost column, $a_{10} = -4 < 0$, so it is not primal feasible. Thus we select x_1 to leave the basis. Among coefficients in the first row, $a_{12} = -3$ and $a_{14} = -1$ are potential pivots. Since

$$\left|\frac{a_{02}}{a_{12}}\right| = \left|\frac{3}{-3}\right| < \left|\frac{a_{04}}{a_{14}}\right| = \left|\frac{5}{-1}\right|$$

we select x_2 to enter the basis. We then make a_{12} equal to 1 by multiplying the first row by $-\left(\frac{1}{3}\right)$. Then, we obtain Tableau 5.2 via row reductions.

In Tableau 5.2, $a_{0j} \geq 0 \quad (j = 1, 2, 3, 4)$ and $a_{i0} \geq 0 \quad (i = 1, 2)$, so the optimum solution is $x_2 = \frac{4}{3}$, $x_3 = \frac{5}{3}$, $x_1 = x_4 = 0$, and $z = 3$.

Tableau 5.2

	1	x_1	x_2	x_3	x_4
$-z$	−3	1	0	0	4
x_2	$\frac{4}{3}$	$-\frac{1}{3}$	1	0	$\frac{1}{3}$
x_3	$\frac{5}{3}$	$\frac{1}{3}$	0	1	$\frac{2}{3}$

The dual of the linear program (5.9) is

$$\begin{aligned} \max \quad & w = -1 - 4\pi_1 + 3\pi_2 \\ \text{subject to} \quad & \pi_1 \leq 0 \\ & -3\pi_1 + \pi_2 \leq 3 \\ & \pi_2 \leq 0 \\ & -\pi_1 + \pi_2 \leq 5 \\ & \pi_i \lesseqgtr 0 \end{aligned} \tag{5.10}$$

The optimum solution of the dual is $\pi_1 = -1$, $\pi_2 = 0$. (This can be solved by first converting it into equations and introducing $\boldsymbol{\pi} = \boldsymbol{y} - \boldsymbol{e}y_0$. Then we can use the

Simplex Method.) The top row of the Simplex Method is $\overline{c}_j = c_j - \boldsymbol{\pi a}_j$, and if $\overline{c}_j \geq 0$, then $\boldsymbol{\pi}$ is a solution to the dual program. Here in the starting Tableau 5.1, $c_1 = c_3 = 0$, $\boldsymbol{a}_1 = [1, 0]$, and $\boldsymbol{a}_3 = [0, 1]$. Therefore what appears in the top row under x_1 and x_3 is $0 - \boldsymbol{\pi e}_j = -\pi_j$.

For example, in Tableau 5.2, we have $1 = -\pi_1$ and $0 = -\pi_2$. Thus, the optimum tableau contains optimum solutions of both primal and dual programs. This being the case, we have the choice of solving either the original program or its dual and the choice of using either the primal or the dual method. In this example, it is unwise to solve the dual program because we would need to add four slack variables, and it would become a program with seven non-negative variables and four equations.

Note
1. $\overline{c}_j \geq 0$ implies that we have a dual feasible solution, since $\overline{c}_j = c_j - \boldsymbol{\pi a}_j \geq 0$, where $\boldsymbol{\pi}$ satisfies the dual constraint.
2. We first decide which variable to leave the basis and then which variable to enter the basis (the opposite of Simplex).
3. Only negative elements are pivot candidates.
4. The "ratio test" is used to maintain dual feasibility.

5.5 Exercises

1. Determine the dual of the given minimization problem:

$$\begin{aligned} \min \quad & z = 3x_1 + 3x_2 \\ \text{subject to} \quad & 2x_1 + x_2 \geq 4 \\ & x_1 + 2x_2 \geq 4 \\ & x_1, x_2 \geq 0 \end{aligned}$$

2. Determine the dual of the given minimization problem:

$$\begin{aligned} \min \quad & z = 4x_1 + x_2 + x_3 \\ \text{subject to} \quad & 3x_1 + 2x_2 + x_3 \geq 23 \\ & x_1 \phantom{{}+2x_2} + x_3 \geq 10 \\ & 8x_1 + x_2 + 2x_3 \geq 40 \\ & x_1, x_2, x_3 \geq 0 \end{aligned}$$

3. Solve the given problem using the Dual Simplex Method and use the Simplex Method to verify optimality:

$$\begin{aligned} \max \quad & z = -5x_1 - 35x_2 - 20x_3 \\ \text{subject to} \quad & x_1 - 2x_2 - x_3 \le -2 \\ & -x_1 - 3x_2 \le -3 \\ & x_1, x_2, x_3 \ge 0 \end{aligned}$$

4. Solve the given problem using the Dual Simplex Method and use the Simplex Method to verify optimality:

$$\begin{aligned} \max \quad & z = -2x_1 - x_2 \\ \text{subject to} \quad & -2x_1 + x_2 + x_3 \le -4 \\ & x_1 + 2x_2 - x_3 \le -6 \\ & x_1, x_2, x_3 \ge 0 \end{aligned}$$

5. Solve the given problem using the Dual Simplex Method and use the Simplex Method to verify optimality:

$$\begin{aligned} \max \quad & z = -4x_1 - 12x_2 - 18x_3 \\ \text{subject to} \quad & x_1 + 3x_2 \ge 3 \\ & 2x_2 + 2x_3 \ge 5 \\ & x_1, x_2, x_3 \ge 0 \end{aligned}$$

6. Solve the given problem using the Dual Simplex Method and use the Simplex Method to verify optimality:

$$\begin{aligned} \min \quad & z = 2x_1 + x_2 \\ \text{subject to} \quad & 3x_1 + 4x_2 \le 24 \\ & 4x_1 + 3x_2 \ge 12 \\ & -x_1 + 2x_2 \ge 1 \\ & x_1 \ge 2, x_2 \ge 0 \end{aligned}$$

Revised Simplex Method

6

In this chapter, we will learn about a method that is mathematically equivalent to the Simplex Method but which can exploit sparsity of the constraint matrix $\boldsymbol{A}$ to run with greater computational efficiency.

6.1 Revised Simplex Method

In the Simplex Method, all entries of the Simplex tableau are changed from one iteration to the next iteration. As such, the Simplex Method requires lots of calculations and storage, and we would like a more efficient method to solve a linear program. Assume that we have an $m \times n$ constraint matrix $\boldsymbol{A}$ and the optimum tableau is obtained in, say, the pth iteration. Then, effectively, we have calculated $p(m+1)(n+1)$ numbers. Notice that once a tableau is obtained during the calculation, we have all the information necessary to calculate the next iteration; all preceding tableaus, including the starting tableau, can be ignored.

Suppose that we keep the starting tableau, and we wish to generate all the entries in a particular tableau. What information is needed? Let us say that we are interested in all the entries in the 29th tableau. Then all we need is $\boldsymbol{B}^{-1}$ associated with the 29th tableau and the names of the current basic variables. All other entries of the 29th tableau can be generated from the entries of the starting tableau and the current $\boldsymbol{B}^{-1}$ of the 29th tableau. Note that $\boldsymbol{\pi} = \boldsymbol{c}_B \boldsymbol{B}^{-1}$, which means the current shadow price $\boldsymbol{\pi}$ is obtained by multiplying $\boldsymbol{c}_B$ of the starting tableau by the current $\boldsymbol{B}^{-1}$.

We define $\overline{\boldsymbol{b}}$ as $\overline{\boldsymbol{b}} = \boldsymbol{B}^{-1}\boldsymbol{b}$, where $\boldsymbol{B}^{-1}$ is from the 29th tableau and $\boldsymbol{b}$ is from the starting tableau. Similarly, we let any column $\bar{\boldsymbol{a}}_j$ be given by $\overline{\boldsymbol{a}}_j = \boldsymbol{B}^{-1}\boldsymbol{a}_j$, where $\boldsymbol{B}^{-1}$ is from the 29th tableau and $\boldsymbol{a}_j$ is from the starting tableau. The *relative cost* (or the *modified cost*) is $\overline{c}_j = c_j - \boldsymbol{\pi}\boldsymbol{a}_j$, where c_j and $\boldsymbol{a}_j$ are from the starting tableau and $\boldsymbol{\pi}$ is the current shadow price.

T.C. Hu, A.B. Kahng, *Linear and Integer Programming Made Easy*,
DOI 10.1007/978-3-319-24001-5_6

The idea is that once we have $\boldsymbol{B}^{-1}$ of the 29th tableau and the labels and names of the basic variables, we can generate all entries if we keep the starting tableau. So, what additional entries in the 29th tableau do we need to get $\boldsymbol{B}^{-1}$ of the 30th tableau?

It turns out that we need the non-basic column of the 29th tableau with a negative relative coefficient and the updated right-hand side (RHS), where $\bar{c}_j = c_j - \boldsymbol{\pi a}_j$, $\bar{\boldsymbol{b}} = \boldsymbol{B}^{-1}\boldsymbol{b}$, $\boldsymbol{\pi} = \boldsymbol{c}_B\boldsymbol{B}^{-1}$, $\boldsymbol{c}_B$ is from the starting tableau, and $\boldsymbol{B}^{-1}$ is the inverse of the 29th tableau.

In matrix notation

$$\begin{bmatrix} 1 & -\boldsymbol{\pi} \\ \boldsymbol{0} & \boldsymbol{B}^{-1} \end{bmatrix} \begin{bmatrix} c_j \\ \boldsymbol{a}_j \end{bmatrix} = \begin{bmatrix} c_j - \boldsymbol{\pi a}_j \\ \boldsymbol{B}^{-1}\boldsymbol{a}_j \end{bmatrix} = \begin{bmatrix} \bar{c}_j \\ \bar{\boldsymbol{a}}_j \end{bmatrix}$$

and

$$\begin{bmatrix} 1 & -\boldsymbol{\pi} \\ \boldsymbol{0} & \boldsymbol{B}^{-1} \end{bmatrix} \begin{bmatrix} 0 \\ \boldsymbol{b} \end{bmatrix} = \begin{bmatrix} -\boldsymbol{\pi b} \\ \boldsymbol{B}^{-1}\boldsymbol{b} \end{bmatrix} = \begin{bmatrix} -\bar{z} \\ \bar{\boldsymbol{b}} \end{bmatrix}$$

In the first equation, if $\bar{c}_j \geq 0$, we are not interested in the components of $\bar{\boldsymbol{a}}_j$. Only if $\bar{c}_j < 0$ do we want to multiply $\boldsymbol{B}^{-1}\boldsymbol{a}_j$, do the feasibility test among positive entries in $\bar{\boldsymbol{a}}_j$ and the right-hand $\bar{\boldsymbol{b}}$, and determine the pivot.

In the starting tableau, the identity is its own inverse, and $\boldsymbol{\pi}$ of the starting tableau is zero because $\boldsymbol{\pi} = \boldsymbol{c}_B\boldsymbol{B}^{-1}$ and $\boldsymbol{c}_B = \boldsymbol{0}$ in the starting tableau. Thus, we can adjoin a column of $[1, 0, \ldots, 0]$ to the identity matrix in the starting tableau, so that we have

$$\begin{bmatrix} 1 & \boldsymbol{0} \\ \boldsymbol{0} & \boldsymbol{I} \end{bmatrix} = \begin{bmatrix} 1 & -\boldsymbol{\pi} \\ \boldsymbol{0} & \boldsymbol{B}^{-1} \end{bmatrix}$$

Let us consider the same example in Chapter 4 which was solved by the Simplex tableau. (See Tableau 4.1.) Now we can do it by the *revised Simplex Method*. The starting tableau is given below in Tableau 6.1.

We have

$$\boldsymbol{B}^* = \begin{bmatrix} 1 & 0 & 0 \\ 0 & 1 & 0 \\ 0 & 0 & 1 \end{bmatrix}$$

$$\begin{aligned} \bar{c}_3 &= c_3 - \boldsymbol{\pi a}_3 = \boldsymbol{B}_0[c_3, \boldsymbol{a}_3] = (1, 0, 0)[2, -2, 3] = 2 \\ \bar{c}_4 &= c_4 - \boldsymbol{\pi a}_4 = (1, 0, 0)[-2, 1, -1] = -2 \\ \bar{c}_5 &= c_5 - \boldsymbol{\pi a}_5 = (1, 0, 0)[-1, 1, 2] = -1 \end{aligned}$$

We use the row vector $(1, 0, 0)$ to multiply the first column $[2, -2, 3]$ and get $\bar{c}_3 = 2$. Since it is positive, we do not care about the components below it. We use the same row vector $(1, 0, 0)$ to multiply the second column vector $[-2, 1, -1]$ and get $\bar{c}_4 = -2$.

Tableau 6.1

	x_0	x_1	x_2	x_3	x_4	x_5	b
x_0	1	0	0	2	−2	−1	0
x_1	0	1	0	−2	1	1	4
x_2	0	0	1	3	−1	2	2

Therefore, we want to bring in $\overline{\boldsymbol{a}}_4^*$, where

$$\overline{\boldsymbol{a}}_4^* = \boldsymbol{B}^{*-1}\boldsymbol{a}_4^* = \begin{bmatrix} 1 & 0 & 0 \\ 0 & 1 & 0 \\ 0 & 0 & 1 \end{bmatrix} \begin{bmatrix} -2 \\ 1 \\ -1 \end{bmatrix} = \begin{bmatrix} -2 \\ 1 \\ -1 \end{bmatrix}$$

$$\overline{\boldsymbol{b}}^* = \boldsymbol{B}^{*-1}\boldsymbol{b}^* = \begin{bmatrix} 1 & 0 & 0 \\ 0 & 1 & 0 \\ 0 & 0 & 1 \end{bmatrix} \begin{bmatrix} 0 \\ 4 \\ 2 \end{bmatrix} = \begin{bmatrix} 0 \\ 4 \\ 2 \end{bmatrix}$$

Since in $\overline{\boldsymbol{a}}_4^*, \overline{a}_{14} = 1$ is the only positive coefficient, we shall drop $\boldsymbol{a}_1$ from the basis. We then perform the pivot operation with $\bar{a}_{14}$ as the pivot (Tableau 6.2). The new tableau is Tableau 6.3.

We put question marks in positions to indicate that they have not been calculated.

Now the inverse of the new basis is

$$\boldsymbol{B}^{*-1} = \begin{bmatrix} 1 & 2 & 0 \\ 0 & 1 & 0 \\ 0 & 1 & 1 \end{bmatrix}$$

$$\overline{c}_1 = (1, 2, 0)[0, 1, 0] = 2$$

$$\overline{c}_3 = (1, 2, 0)[2, -2, 3] = -2$$

$$\overline{c}_5 = (1, 2, 0)[-1, 1, 2] = 1$$

As a check, we have

$$\overline{c}_2 = (1, 2, 0)[0, 0, 1] = 0$$

$$\overline{c}_4 = (1, 2, 0)[-2, 1, -2] = 0$$

We use the row vector $(1, 2, 0)$ to multiply the column vector in the starting tableau $[-1, 1, 2]$ and get $\overline{c}_5 = 1$, so we do not care about the rest of the entries. Going back, we use the same row vector $(1, 2, 0)$ to multiply the column vector $[2, -2, 3]$ and get $\overline{c}_3 = -2$.

Tableau 6.2

	x_0	x_1	x_2	x_3	x_4	x_5	b
x_0	1	0	0	2	−2	1	0
x_1	0	1	0	?	1*	?	4
x_2	0	0	1	?	−1	?	2

Tableau 6.3

	x_0	x_1	x_2	x_3	x_4	x_5	b
x_0	1	2	0	?	0	?	8
x_4	0	1	0	?	1	?	4
x_2	0	1	1	?	0	?	6

Therefore, we should bring $\boldsymbol{a}_3$ into the basis.

$$\overline{\boldsymbol{a}}_3^* = \boldsymbol{B}^{*-1}\boldsymbol{a}_3^* = \begin{bmatrix} 1 & 2 & 0 \\ 0 & 1 & 0 \\ 0 & 1 & 1 \end{bmatrix} \begin{bmatrix} 2 \\ -2 \\ 3 \end{bmatrix} = \begin{bmatrix} -2 \\ -2 \\ 1 \end{bmatrix}$$

$$\overline{\boldsymbol{b}}^* = \boldsymbol{B}^{*-1}\boldsymbol{b}^* = \begin{bmatrix} 1 & 2 & 0 \\ 0 & 1 & 0 \\ 0 & 1 & 1 \end{bmatrix} \begin{bmatrix} 0 \\ 4 \\ 2 \end{bmatrix} = \begin{bmatrix} 8 \\ 4 \\ 6 \end{bmatrix}$$

Since $\overline{a}_{23} = 1$ is the only positive element, $\boldsymbol{a}_2$ should be dropped. We perform the pivot operation with $\overline{a}_{23}$ as the pivot (Tableau 6.4). The new tableau is Tableau 6.5.

Now the inverse of the new basis is

$$\boldsymbol{B}^{*-1} = \begin{bmatrix} 1 & 4 & 2 \\ 0 & 3 & 2 \\ 0 & 1 & 1 \end{bmatrix}$$

$$\overline{c}_1 = (1,4,2)[0,1,0] = 4$$
$$\overline{c}_2 = (1,4,2)[0,0,1] = 2$$
$$\overline{c}_5 = (1,4,2)[-1,1,2] = 7$$

As a check, we have

$$\overline{c}_2 = (1,4,2)[2,-2,3] = 0$$
$$\overline{c}_3 = (1,4,2)[-2,1,-1] = 0$$

Finally, we have Tableau 6.6. We have as an optimum solution $x_4 = 16$, $x_3 = 6$, $x_1 = x_2 = x_5 = 0$, and $x_0 = 20$. Note that in the revised Simplex Method, we do not calculate all $\overline{c}_j$. Once some $\overline{c}_j < 0$ is found, we can immediately bring the column into the basis.

Tableau 6.4

	x_0	x_1	x_2	x_3	x_4	x_5	b
x_0	1	2	0	−2	0	1	8
x_4	0	1	0	−2	?	?	4
x_2	0	1	1	1*	?	?	6

Tableau 6.5

	x_0	x_1	x_2	x_3	x_4	x_5	b
x_1	1	4	2	0	?	?	20
x_4	0	3	2	0	?	?	16
x_3	0	1	1	1	?	?	6

Tableau 6.6

	x_0	x_1	x_2	x_3	x_4	x_5	b
x_0	1	4	2	0	0	7	20
x_4	0	3	2	?	?	?	16
x_3	0	1	1	?	?	?	6

Consider a linear program

$$\begin{aligned} \min \quad & z = c_1x_1 + c_2x_2 + c_3x_3 \\ \text{subject to} \quad & a_{11}x_1 + a_{12}x_2 + a_{13}x_3 = b_1 \\ & a_{21}x_1 + a_{22}x_2 + a_{23}x_3 = b_2 \\ & x_j \geq 0 \end{aligned} \tag{6.1}$$

We can find scalars for each row, say π_1 for the first row and π_2 for the second, such that the columns of the constraint equations sum to equal the corresponding c_i. In other words,

$$\begin{aligned} c_1 - \pi_1 a_{11} - \pi_2 a_{21} = 0 \\ c_2 - \pi_1 a_{12} - \pi_2 a_{22} = 0 \end{aligned}$$

For example, take the linear program

$$\begin{aligned} \min \quad & z = x_1 + x_2 + x_3 \\ \text{subject to} \quad & x_1 - x_2 + 2x_3 = 2 \\ & -x_1 + 2x_2 - x_3 = 1 \end{aligned} \tag{6.2}$$

If we multiply the first constraint equation by $\pi_1 = 3$ and the second by $\pi_2 = 2$ and subtract from the objective function, we obtain

$$1-3(1)-2(-1)=0$$
$$1-3(-1)-2(2)=0$$

Notice that

$$\begin{bmatrix}\pi_1 & \pi_2\end{bmatrix} = \begin{bmatrix}c_1 & c_2\end{bmatrix}\begin{bmatrix}a_{11} & a_{12} \\ a_{21} & a_{22}\end{bmatrix}^{-1}$$

After we put the linear program into diagonal form with respect to $-z, x_1, x_2$, we obtain the equations

$$\begin{aligned}
-z \qquad\qquad &+ \bar{c}_3 x_3 = -\bar{z}_0 \\
x_1 \qquad &+ \bar{a}_{13} x_3 = \bar{b}_1 \\
x_2 &+ \bar{a}_{23} x_3 = \bar{b}_2
\end{aligned}$$

which, in our example, correspond to

$$\begin{aligned}
-z \qquad\qquad &- 3x_3 = -8 \\
x_1 \qquad &+ 3x_3 = 5 \\
x_2 &+ \ \ x_3 = 3 \\
x_j \geq 0 &
\end{aligned} \tag{6.3}$$

Here, we have a feasible solution with basic variables $x_1 = 5, x_2 = 3$ and non-basic variable $x_3 = 0$. If the value of the non-basic variable is increased from zero, then the value of the objective function will decrease.

The maximum value by which x_3 can be increased is limited by the ratio test

$$x_3 = \min\left[\frac{5}{3}, \frac{3}{1}\right] = \frac{5}{3}$$

because the pivot is a_{13}, which means that x_1 will become a non-basic variable and x_3 will become a basic variable.

If we put the linear program into a diagonal form with respect to $-z, x_2, x_3$, we obtain

$$\begin{aligned}
-z + x_1 \qquad\qquad &= -3 \\
-\frac{1}{3}x_1 + x_2 \qquad &= \frac{4}{3} \\
\frac{1}{3}x_1 \qquad + x_3 &= \frac{5}{3}
\end{aligned} \tag{6.4}$$

In other words,

$$\begin{aligned}
-z + \bar{c}_1 x_1 \qquad\qquad &= -3 \\
\bar{a}_{21} x_1 + x_2 \qquad &= \frac{4}{3} \\
\bar{a}_{11} x_1 \qquad + x_3 &= \frac{5}{3}
\end{aligned}$$

where $\bar{c}_j \geq 0$ for all non-basic variables. This condition indicates that the optimum solution has been obtained. Note that (6.4) is obtained from (6.3) by multiplying the first row by -1 and the second row by 0 and subtracting from the 0th row.

In general, after a linear program is in diagonal form, such as (6.4), whatever appears above the identity matrix $\boldsymbol{I}$ in the 0th row is the current $(-\pi_1, -\pi_2, \ldots, -\pi_m)$. Also, whatever appears in the position of $\boldsymbol{I}$ is the current $\boldsymbol{B}^{-1}$. For example, the following matrix occupies the position of $\boldsymbol{I}$ in (6.4):

$$\begin{bmatrix} \frac{1}{3} & 0 \\ -\frac{1}{3} & 1 \end{bmatrix} \begin{bmatrix} 3 & 0 \\ 1 & 1 \end{bmatrix} = \begin{bmatrix} 1 & 0 \\ 0 & 1 \end{bmatrix}$$

If we interpret the example linear program (6.2) as the task of buying adequately nutritional food as cheaply as possible, then there are three kinds of food, each costing one dollar per unit $(c_1 = c_2 = c_3 = 1)$. The first food contains one unit of vitamin A and destroys one unit of vitamin B, and the second food destroys one unit of vitamin A and contains two units of vitamin B.

If we decide to buy five units of the first food and three units of the second food, the minimum requirements will be satisfied at the cost of eight dollars.

$$5\begin{bmatrix} 1 \\ -1 \end{bmatrix} + 3\begin{bmatrix} -1 \\ 2 \end{bmatrix} = \begin{bmatrix} 2 \\ 1 \end{bmatrix}$$

This is equivalent to paying three dollars for one unit of vitamin A and two dollars for one unit of vitamin B, since

$$2\begin{bmatrix} 1 \\ 0 \end{bmatrix} + 1\begin{bmatrix} 0 \\ 1 \end{bmatrix} = \begin{bmatrix} 2 \\ 1 \end{bmatrix}$$

and $3x_1 + 2x_2 = 8$ dollars.

In general, once we decide to select a set of items to buy (i.e., select a set of columns to be the basic columns), then there exists a set of shadow prices $(\pi_1, \pi_2, \ldots, \pi_m)$, one for each row, such that

$$\boldsymbol{\pi} = \boldsymbol{c}_B \boldsymbol{B}^{-1} \tag{6.5}$$

6.2 Simplified Calculation

Consider a linear program in matrix form

$$\begin{aligned} \min \quad & z = \boldsymbol{cx} \\ \text{subject to} \quad & \boldsymbol{Ax} = \boldsymbol{b} \\ & \boldsymbol{x} \geq \boldsymbol{0} \end{aligned} \tag{6.6}$$

Let us partition $\boldsymbol{c}$ into $(\boldsymbol{c}_B, \boldsymbol{c}_N)$, $\boldsymbol{A}$ into $(\boldsymbol{B}, \boldsymbol{N})$, and $\boldsymbol{x}$ into $[\boldsymbol{x}_B, \boldsymbol{x}_N]$. Then we can rewrite the linear program as

$$\begin{aligned} \min \quad & z = \boldsymbol{c}_B \boldsymbol{x}_B + \boldsymbol{c}_N \boldsymbol{x}_N \\ \text{subject to} \quad & \boldsymbol{B}\boldsymbol{x}_B + \boldsymbol{N}\boldsymbol{x}_N = \boldsymbol{b} \\ & \boldsymbol{x}_B, \boldsymbol{x}_N \geq \boldsymbol{0} \end{aligned} \tag{6.7}$$

If we select $\boldsymbol{B}$ to be the basis, we multiply the constraints by $\boldsymbol{B}^{-1}$ and obtain

$$\boldsymbol{I}\boldsymbol{x}_B + \boldsymbol{B}^{-1}\boldsymbol{N}\boldsymbol{x}_N = \boldsymbol{B}^{-1}\boldsymbol{b}$$

If we multiply (6.7) by $\boldsymbol{\pi}$ and subtract from the objective function, we obtain the following:

$$\begin{aligned} -z + (\boldsymbol{c}_B - \boldsymbol{\pi}\boldsymbol{B})\boldsymbol{x}_B + (\boldsymbol{c}_N - \boldsymbol{\pi}\boldsymbol{N})\boldsymbol{x}_N &= -\boldsymbol{\pi}\boldsymbol{b} \\ -z + \quad 0 \quad + (\boldsymbol{c}_N - \boldsymbol{\pi}\boldsymbol{N})\boldsymbol{x}_N &= -\boldsymbol{\pi}\boldsymbol{b} \\ z \quad &= \boldsymbol{\pi}\boldsymbol{b} \end{aligned}$$

In other words, z equals the original RHS $\boldsymbol{b}$ multiplied by the current shadow prices. In the Simplex Method, every entry in

$$\begin{bmatrix} \boldsymbol{c} & 0 \\ \boldsymbol{A} & \boldsymbol{b} \end{bmatrix} = \begin{bmatrix} \boldsymbol{c}_B & \boldsymbol{c}_N & 0 \\ \boldsymbol{B} & \boldsymbol{N} & \boldsymbol{b} \end{bmatrix}$$

is changed from iteration to iteration; for example, the 100th tableau is obtained from the 99th tableau. If we knew the value of $\boldsymbol{B}^{-1}$ for the 99th tableau, we could directly generate every entry in the 100th tableau by multiplying the original tableau by $\boldsymbol{B}^{-1}$ of the 99th tableau without going through the first 99 iterations.

Since we do not know $\boldsymbol{B}^{-1}$ of the 99th tableau, we cannot go directly from the first tableau to the 100th. But we can save work by keeping *only the value of* $\boldsymbol{B}^{-1}$ *for every tableau*, together with the original tableau. The size of $\boldsymbol{B}^{-1}$ is only $(m+1) \times (m+1)$, which may be much smaller than $(m+1) \times n$. This technique works because we are only interested in a non-basic column j when $\bar{c}_j < 0$. If all $\bar{c}_j \geq 0$, then we are not interested in generating the matrix $\boldsymbol{N}$ corresponding to the non-basic columns. In the worst case, we only need to generate the following portions of the tableau:

	$+ \; + \; + \; \cdots \; +$	$-$	
$\boldsymbol{B}^{-1}$		$\pm$ $\pm$ $\pm$ $\pm$ $\pm$	$\boldsymbol{b}$

(Note that $\pm$ denotes an unrestricted variable.)

Note that if $\bar{c}_j = c_j - \boldsymbol{\pi}\boldsymbol{a}_j \geq 0$ for all j, then the current solution is optimal.

Column Generating Technique 7

In this chapter, we shall discuss the most important computational technique of solving a large linear program. If a linear program has a very large number of columns (variables), perhaps too many to write down, we can use the *column generating* technique to obtain the optimum solution. The point is that entries of the constraint matrix are not random numbers. For every linear program, there is a special structure which is implicitly given in the matrix. By exploring this special structure, we can solve a large linear program with infinitely many columns.

Since we cannot write down all columns of the constraint matrix, how can we be sure that all the modified costs (= relative cost, in Chapter 6) are non-negative, i.e., $\overline{c}_j = c_j - \boldsymbol{\pi a}_j \geq 0$? It turns out that if the minimum of all modified costs is non-negative, then all modified costs are non-negative. How can we be sure that the minimum modified cost is non-negative? Well, we need to solve an optimization problem to find the column with the most negative modified cost.

If the least modified cost is indeed negative, then we shall let that column enter the current basis to replace another column. An overview of the column generating technique is shown in Figure 7.1. The iteration process stops when the most negative modified cost (relative cost) is non-negative.

Given a feasible solution of the large linear program, let the shadow price be $\boldsymbol{\pi}$. If the coefficient c_j in the linear program is the same, then we try to find the column $\boldsymbol{a}_j$ which leads to the maximum $\boldsymbol{\pi a}_j$. Thus, we can find the minimum relative cost since the relative cost $\overline{c}_j$ is equal to $c_j - \boldsymbol{\pi a}_j$.

Intuitively speaking, we want the column $\boldsymbol{a}_j$ which has a small coefficient c_j, the product with the shadow prices of the current basis to be large, and the most compatibility with $m - 1$ columns of the current basis.

T.C. Hu, A.B. Kahng, *Linear and Integer Programming Made Easy*,
DOI 10.1007/978-3-319-24001-5_7

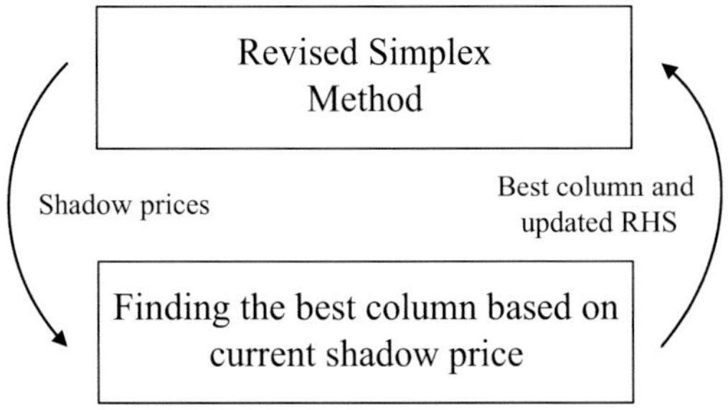

Fig. 7.1 The overview of the column generating technique

Let us illustrate the idea in a numerical example:

$$\begin{aligned} \min \quad & z = x_1 + x_2 + x_3 + x_4 + x_5 \\ \text{subject to} \quad & \begin{bmatrix} 2 \\ 4 \end{bmatrix} x_1 + \begin{bmatrix} 5 \\ 1 \end{bmatrix} x_2 + \begin{bmatrix} 4 \\ 3 \end{bmatrix} x_3 + \begin{bmatrix} 5 \\ 4 \end{bmatrix} x_4 + \begin{bmatrix} 4 \\ 5 \end{bmatrix} x_5 = \begin{bmatrix} 11 \\ 11 \end{bmatrix} \\ & x_j \geq 0 \end{aligned} \tag{7.1}$$

The five columns are plotted as shown in Figure 7.2. The requirement $\begin{bmatrix} 11 \\ 11 \end{bmatrix}$ is shown as a line from the origin to the coordinate $\begin{bmatrix} 11 \\ 11 \end{bmatrix}$. There are four line segments connecting $\begin{bmatrix} 2 \\ 4 \end{bmatrix}$ to $\begin{bmatrix} 5 \\ 1 \end{bmatrix}$, $\begin{bmatrix} 2 \\ 4 \end{bmatrix}$ to $\begin{bmatrix} 4 \\ 3 \end{bmatrix}$, $\begin{bmatrix} 2 \\ 4 \end{bmatrix}$ to $\begin{bmatrix} 5 \\ 4 \end{bmatrix}$, and $\begin{bmatrix} 5 \\ 4 \end{bmatrix}$ to $\begin{bmatrix} 4 \\ 5 \end{bmatrix}$.

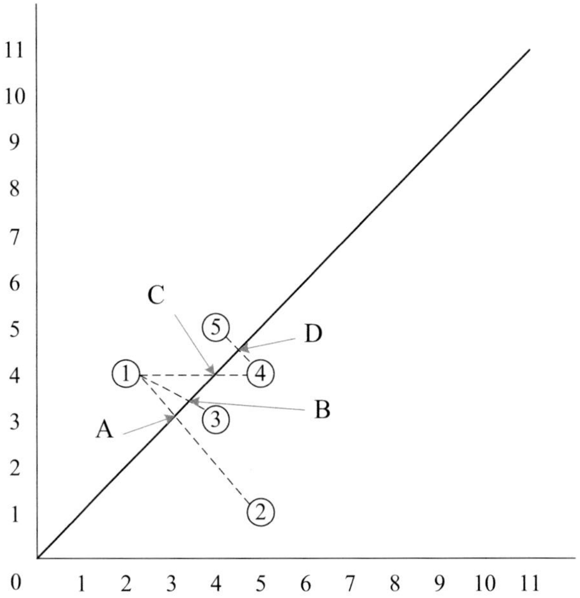

Fig. 7.2 Plotting of the five columns

The intersections of the four dotted line segments with the line from the origin to $\begin{bmatrix} 11 \\ 11 \end{bmatrix}$ are four possible solutions with the values of $z = \frac{11}{3}, \frac{33}{10}, \frac{33}{12}, \frac{22}{9}$. The detailed computations are shown in Tableaus 7.1, 7.2, 7.3, 7.4, 7.5, 7.6, and 7.7.

Tableau 7.1

$-z$	x_1	x_2	x_3	x_4	x_5		*RHS*
1	1	1	1	1	1	...	0
0	2	5	4	5	4	...	11
0	4	1	3	4	5	...	11

The result of multiplying Tableau 7.1 by the inverse of $\begin{bmatrix} 2 & 5 \\ 4 & 1 \end{bmatrix}$ is shown in Tableau 7.2.

Tableau 7.2

$-z$	x_1	x_2	x_3	x_4	x_5		*RHS*
1	1	1	1	1	1	...	0
0	1	0	$\frac{11}{18}$	$\frac{15}{18}$	$\frac{21}{18}$	...	$\frac{44}{18}$
0	0	1	$\frac{10}{18}$	$\frac{12}{18}$	$\frac{6}{18}$	...	$\frac{22}{18}$

The result of subtracting the first row and the second row from the 0th row in Tableau 7.2 is shown in Tableau 7.3.

Tableau 7.3

	$-z$	x_1	x_2	x_3	x_4	x_5		*RHS*
	1	0	0	$-\frac{3}{18}$	$-\frac{9}{18}$	$-\frac{9}{18}$	...	$-\frac{66}{18}$
x_1	0	1	0	$\frac{11}{18}$	$\frac{15}{18}$	$\frac{21}{18}$	...	$\frac{44}{18}$
x_2	0	0	1	$\frac{10}{18}$	$\frac{12}{18}$	$\frac{6}{18}$	...	$\frac{22}{18}$

We have $x_1 = \frac{44}{18}$, $x_2 = \frac{22}{18}$, and $z = \frac{11}{3}$ (A in Figure 7.2).

The result of multiplying the Tableau 7.1 by the inverse of $\begin{bmatrix} 2 & 4 \\ 4 & 3 \end{bmatrix}$ is shown in Tableau 7.4.

Tableau 7.4

$-z$	x_1	x_2	x_3	x_4	x_5		RHS
1	1	1	1	1	1	...	0
0	1	$-\frac{11}{10}$	0	$\frac{1}{10}$	$\frac{8}{10}$	...	$\frac{11}{10}$
0	0	$\frac{18}{10}$	1	$\frac{12}{10}$	$\frac{6}{10}$	...	$\frac{22}{10}$

The result of subtracting the first row and the second row from the 0th row in Tableau 7.4 is shown in Tableau 7.5.

Tableau 7.5

	$-z$	x_1	x_2	x_3	x_4	x_5		RHS
	1	0	$\frac{3}{10}$	0	$-\frac{3}{10}$	$-\frac{4}{10}$	...	$-\frac{33}{10}$
x_1	0	1	$-\frac{11}{10}$	0	$\frac{1}{10}$	$\frac{8}{10}$	...	$\frac{11}{10}$
x_3	0	0	$\frac{18}{10}$	1	$\frac{12}{10}$	$\frac{6}{10}$	...	$\frac{22}{10}$

We have $x_1 = \frac{11}{10}$, $x_3 = \frac{22}{10}$, and $z = \frac{33}{10}$ (B in Figure 7.2).

The result of multiplying Tableau 7.1 by the inverse of $\begin{bmatrix} 2 & 5 \\ 4 & 4 \end{bmatrix}$ is shown in Tableau 7.6.

Tableau 7.6

$-z$	x_1	x_2	x_3	x_4	x_5		RHS
1	1	1	1	1	1	...	0
0	1	$-\frac{15}{12}$	$-\frac{1}{12}$	0	$\frac{9}{12}$	...	$\frac{11}{12}$
0	0	$\frac{18}{12}$	$\frac{10}{12}$	1	$\frac{6}{12}$	...	$\frac{22}{12}$

The result of subtracting the first row and the second row from the 0th row in Tableau 7.6 is shown in Tableau 7.7.

Tableau 7.7

	$-z$	x_1	x_2	x_3	x_4	x_5		RHS
	1	0	$\frac{9}{12}$	$\frac{3}{12}$	0	$-\frac{3}{12}$	...	$-\frac{33}{12}$
x_1	0	1	$-\frac{15}{12}$	$\frac{-1}{12}$	0	$\frac{9}{12}$	...	$\frac{11}{12}$
x_4	0	0	$\frac{18}{12}$	$\frac{10}{12}$	1	$\frac{6}{12}$	...	$\frac{22}{12}$

We have $x_1 = \frac{11}{12}$, $x_4 = \frac{22}{12}$, and $z = \frac{33}{12}$ (C in Figure 7.2).

The result of multiplying Tableau 7.1 by the inverse of $\begin{bmatrix} 5 & 4 \\ 4 & 5 \end{bmatrix}$ is shown in Tableau 7.8.

Tableau 7.8

$-z$	x_1	x_2	x_3	x_4	x_5		RHS
1	1	1	1	1	1	...	0
0	$-\frac{6}{9}$	$\frac{21}{9}$	$\frac{8}{9}$	1	0	...	$\frac{11}{9}$
0	$\frac{12}{9}$	$-\frac{15}{9}$	$-\frac{1}{9}$	0	1	...	$\frac{11}{9}$

The result of subtracting the first row and the second row in Tableau 7.8 from the 0th row is shown in Tableau 7.9.

Tableau 7.9

	$-z$	x_1	x_2	x_3	x_4	x_5		RHS
	1	$\frac{3}{9}$	$\frac{3}{9}$	$\frac{2}{9}$	0	0	...	$-\frac{22}{9}$
x_4	0	$-\frac{6}{9}$	$\frac{21}{9}$	$\frac{8}{9}$	1	0	...	$\frac{11}{9}$
x_5	0	$\frac{12}{9}$	$-\frac{15}{9}$	$-\frac{1}{9}$	0	1	...	$\frac{11}{9}$

We have $x_4 = \frac{11}{9}$, $x_5 = \frac{11}{9}$, and $z = \frac{22}{9}$ (D in Figure 7.2).

In the numerical example, all c_j are equal to one, so the modified cost $\overline{c}_j = c_j - \boldsymbol{\pi a}_j$ is minimum if $\boldsymbol{\pi a}_j$ is maximum.

The first shadow price $\boldsymbol{\pi} = \left(\frac{1}{6}, \frac{1}{6}\right)$ since $\boldsymbol{c}_B \cdot \boldsymbol{B}^{-1} = (1, 1) \cdot \begin{bmatrix} 2 & 5 \\ 4 & 1 \end{bmatrix}^{-1} = \boldsymbol{\pi}$. So, we look for an $\boldsymbol{a}_j$ which has $\max_j \left(a_{1j} + a_{2j}\right)$.

Consider the Homemaker and Pill Salesperson examples in Chapter 2. Let the supermarket have only two kinds of food: one kind has two units of vitamin A and four units of vitamin B, and the other kind has five units of vitamin A and one unit of vitamin B. Furthermore, let the homemaker's family need 11 units of vitamin A and 11 units of vitamin B. Then the homemaker will pay $\frac{66}{18}$ dollars. In the meantime, the pill salesperson will charge $\frac{1}{6}$ dollars per capsule for vitamin A and $\frac{1}{6}$ dollars per capsule for vitamin B,

$$11\left(\frac{1}{6}+\frac{1}{6}\right)=\frac{66}{18}$$

Remember $\boldsymbol{\pi} \cdot \boldsymbol{b} = z$. In the example,

$$\boldsymbol{\pi} \cdot \boldsymbol{b} = \left(\frac{1}{6}, \frac{1}{6}\right) \cdot \begin{bmatrix} 11 \\ 11 \end{bmatrix} = \frac{66}{18}$$

Once the supermarket has a third kind of food with vitamin contents $[4, 3]$ for vitamins A and B, the homemaker will buy the foods $[2, 4]$ and $[4, 3]$, and the pill salesperson will decrease the price of vitamin A capsule to $\frac{1}{10}$ dollars and increase the price of vitamin B capsule to $\frac{1}{5}$ dollars.

Let us look at Figure 7.2 again. When the shadow price $\boldsymbol{\pi} = (\pi_1, \pi_2) = \left(\frac{1}{6}, \frac{1}{6}\right)$, the best vector is one with $\max_j (a_{1j} = a_{2j})$, with three kinds of food having negative modified cost. When the shadow price becomes $(\pi_1, \pi_2) = \left(\frac{1}{9}, \frac{1}{9}\right)$, i.e., $z = \frac{22}{9}$, then all five kinds of food have non-negative modified costs (see the 0th column of Tableau 7.9), and the computation stops.

The following is a brief summary of the Simplex Method, the revised Simplex Method, and the column generating technique.

1. Simplex Method: Search the top row and use the most negative column to replace a column in the current basis until $\bar{c}_j \geq 0$ for all columns. (Too much storage and work.)
2. Revised Simplex Method: Keep a tableau of size $(m+1) \times (n+2)$. Use a column to replace another in the current basis if $\bar{c}_j < 0$ (where $\bar{c}_j = c_j - \boldsymbol{\pi a}_j$). Only generate the entries of the column when $\bar{c}_j < 0$. Use a ratio test to keep the next solution feasible. Since the exchange uses one new column and replaces one old column, the work is reasonable.

   Notice that the result of $z = \boldsymbol{\pi b}$ can be checked in each iteration, where $\boldsymbol{\pi}$ is the current shadow price and $\boldsymbol{b}$ is the original right-hand side.
3. Column Generating Technique: This technique is typically applied to linear programs with too many columns to write down. The coefficients in the Simplex tableau are not random numbers. The entries in a column could be the path from one point to another point, or perhaps they could be weights of several items in a knapsack packing problem. The columns could be partitioned into several sets, or each set of columns could be the multicommodity flows of a special commodity. Therefore, the problem of selecting a column to enter the basis is an optimization problem itself. The optimum solution is found when the best column has $\bar{c}_j \geq 0$.

The Knapsack Problem

8

We now turn to the subject of *integer linear programming*, or *integer programming* for short. Linear programming, which permits non-integer values of the solution variables, is known to be "easy" in the sense of having known algorithms with worst-case runtime that is polynomial in the size of the problem instance. By contrast, integer programming (as well as the hybrid known as mixed integer linear programming) requires that at least some solution variables have their values restricted to be integers. The problem then becomes "hard": there is no known algorithm that guarantees a worst-case runtime that is polynomial in the instance size.

Since so many important optimizations can be framed as integer programs, you can find many textbooks that cover a vast range of methods for integer programming. For example, *branch and bound* methods systematically traverse a tree of all possible solutions, avoiding exploration of subtrees that have infeasible solutions. On the other hand, *cutting plane* methods start with a continuous, linear program "relaxation" of the integer program but then gradually enforce the integer constraint. This book's website gives some links to references on such techniques.

In this book, we highlight the asymptotic behavior of integer programs as their right-hand sides become large. In practice, this asymptotic behavior can make certain integer programs more tractable to efficient solution. For now, we start with the simplest integer program, called the *knapsack problem*.

8.1 Introduction

We are now going to be looking at integer programs. We start with the simplest integer program, which has a single constraint. Consider a hiker with a knapsack that can carry 20 lbs. There are three kinds of items with various values and

T.C. Hu, A.B. Kahng, *Linear and Integer Programming Made Easy*,
DOI 10.1007/978-3-319-24001-5_8

weights, and he wants to select items to put in the knapsack with maximum total value. So we have the *knapsack problem*, as follows:

$$\begin{aligned} \max \quad & v = 12x_1 + 10x_2 + 7x_3 \\ \text{subject to} \quad & 11x_1 + 10x_2 + 9x_3 \le 20 \\ & x_j \ge 0 \quad \text{integers} \end{aligned} \tag{8.1}$$

If we use the Ratio Method, the first item has the best ratio, with

$$\frac{12}{11} > \frac{10}{10} > \frac{7}{9}$$

Since $x_1 = \frac{20}{11}$ is not an integer, $x_1 = \left\lfloor \frac{20}{11} \right\rfloor = 1$. So the integer program (8.1) would give $x_1 = 1$ and $x_3 = 1$ with a total value of 19 dollars. However, the optimum integer solution of (8.1) is $x_2 = 2$ with a total value of 20 dollars. Here, we can see the difficulty of integer programs in comparison to linear programs.

Through this simple example, we can see some features of integer programs:

1. The best item may not be used in the optimum integer solution.
2. The maximum linear program solution, $x_1 = \frac{20}{11}$ with objective function value $\frac{20}{11} \cdot 12 = \frac{240}{11} \cong 21.82$, is usually better than the optimum integer solution (the integer constraint on variables certainly cannot help the optimum solution's value).
3. We cannot round off the optimum linear program solution to get the optimum integer solution, and the difference between the two values can be arbitrarily large. For example, we can change (8.1) to (8.2):

$$\begin{aligned} \max \quad & v = 1002x_1 + 1000x_2 + x_3 \\ \text{subject to} \quad & 1001x_1 + 1000x_2 + 999x_3 \le 2000 \\ & x_j \ge 0 \quad \text{integers} \end{aligned} \tag{8.2}$$

Consider the following example:

$$\begin{aligned} \max \quad & v = 12x_1 + 10x_2 + \ x_3 \\ \text{subject to} \quad & 11x_1 + 10x_2 + 9x_3 \le 110 \\ & x_j \ge 0 \quad \text{integers} \end{aligned} \tag{8.3}$$

where $v_1 = 12$, $v_2 = 10$, $v_3 = 1$, $w_1 = 11$, $w_2 = 10$, $w_3 = 9$, and $b = 110$. Notice that the right-hand side b equals the least common multiple, $\text{lcm}(w_1, w_2) = 110$. In this case, instead of using $x_2 = \frac{110}{10} = 11$, we could use $x_1 = \frac{110}{11} = 10$ and get more value.

Denote the density of the best item by $p_1 = \frac{v_1}{w_1}$ and the density of the second best item by $p_2 = \frac{v_2}{w_2}$, where p_j denotes the value per pound of the item j. Intuitively, we would fill the knapsack with the first item and then fill leftover space with the second item, etc. This intuitive approach usually gives very good results, but it does not always give the optimum solution due to integer restrictions.

One thing is clear: if b is sufficiently large compared to w_1, then $x_1 > 0$ in the optimum solution. That is, the best item would be used at least once. To see this, let us assume that $x_1 = 0$ in the optimum solution. Then, the maximum value will certainly not exceed $p_2 b$. In other words, the maximum value without the best item (i.e., $x_1 = 0$) is $p_2 b$.

However, if we try to fill the knapsack with only the best item, we can obtain the value

$$v_1 \left\lfloor \frac{b}{w_1} \right\rfloor > v_1 \cdot \left(\frac{b}{w_1} - \frac{w_1}{w_1} \right) = v_1 \cdot \frac{b - w_1}{w_1} = p_1 w_1 \cdot \frac{b - w_1}{w_1} = p_1 (b - w_1)$$

To find the value b which yields

$$v_1 \left\lfloor \frac{b}{w_1} \right\rfloor > p_1 (b - w_1) \geq p_2 b$$

we solve the equation

$$p_1 (b - w_1) = p_2 b \quad \text{or} \quad b = \frac{p_1}{p_1 - p_2} \cdot w_1$$

That is, for $b \geq \dfrac{p_1}{p_1 - p_2} \cdot w_1, \quad p_1 (b - w_1) \geq p_2 b.$ (8.4)

In other words, if b is sufficiently large, $x_1 > 0$ in the optimum solution. Note that the above condition (8.4) and $b \geq \text{lcm}(w_1, w_2)$ are both sufficient conditions but not necessary conditions.

Define

$$F_k(y) = \sum_{j=1}^{k} v_j x_j \quad (0 \leq k \leq n)$$

where

$$\sum_{j=1}^{k} w_j x_j \leq y \quad (0 \leq y \leq b)$$

Here, $F_k(y)$ is the maximum value obtained by using the first k items only when the total weight limitation is y. Notice that $F_k(y)$ can be calculated recursively from $k = 0, 1, \ldots, n$ and $y = 0, 1, \ldots, b$. $F_k(0) = 0$ if the weight-carrying capacity is zero.

If we only use the first item,

$$F_1(y) = v_1 \left\lfloor \frac{y}{w_1} \right\rfloor$$

When there are two kinds of items available, we could either only use the first item or we can use the second item at least once. So,

$$F_2(y) = \max\{F_1(y), \quad v_2 + F_2(y - w_2)\}$$

Note that the formula does not restrict $x_2 = 1$ since the term $F_2(y - w_2)$ could contain $x_2 \geq 1$. Similarly,

$$\begin{aligned} F_3(y) &= \max\{F_2(y), \quad v_3 + F_3(y - w_3)\} \\ F_4(y) &= \max\{F_3(y), \quad v_4 + F_4(y - w_4)\} \\ &\vdots \end{aligned}$$

For a general $F_k(y)$, we have the following recursive relation:

$$F_k(y) = \max\{F_{k-1}(y), \quad v_k + F_k(y - w_k)\} \tag{8.5}$$

When the first k items are chosen to obtain $F_k(y)$, either the kth item is used at least once, or it is not used at all. If it is used at least once, then the total weight limitation is reduced to $y - w_k$. If it is not used, then $F_k(y)$ is the same as $F_{k-1}(y)$.

Remember that $x_1 > 0$ in the optimum solution when b is sufficiently large. This means that

$$F_n(b) = v_1 + F_n(b - w_1) \quad \text{for} \quad b > \frac{p_1 w_1}{p_1 - p_2} \tag{8.6}$$

The most interesting feature of the knapsack problem is that its optimum integer solution will be periodic if the right-hand side bound b becomes asymptotically large.

Define $\theta(b) = p_1 b - F_n(b)$ to be the difference between the optimum linear program value and the optimum integer program value using n kinds of items. Then,

$$\begin{aligned} \theta(b - w_1) &= p_1(b - w_1) - F_n(b - w_1) \\ &= p_1 b - p_1 w_1 - (F_n(b) - v_1) \quad \text{(from (8.6) and for sufficiently large } b\text{)} \\ &= p_1 b - F_n(b) \\ &= \theta(b) \end{aligned}$$

This shows that the function $\theta(b)$ is periodic in nature, with period w_1 for sufficiently large b.

The function $\theta(b)$ can be calculated by

$$\theta(b) = \min_j \{\theta(b - w_j) + (p_1 - p_j)w_j\} \tag{8.7}$$

If $x_j > 0$ in the optimum solution, then $\theta(b) = \theta(b - w_j)$ plus the loss, $(p_1 - p_j) \cdot w_1$, from not fulfilling the weight w_j with the first item. Since there must be some $x_j > 0$ in the optimum solution, $\theta(b)$ is obtained by minimizing with respect to j.

To show the periodic nature of the optimum integer solution, consider the numerical example:

$$\begin{aligned} \max \quad & v = 18x_1 + 14x_2 + 8x_3 + 4x_4 + 0x_5 \\ \text{subject to} \quad & 15x_1 + 12x_2 + 7x_3 + 4x_4 + 1x_5 \le b \\ & x_j \ge 0 \quad \text{integers} \end{aligned}$$

According to (8.4) with $p_1 = \frac{18}{15}$ and $p_2 = \frac{14}{12}$, we have $b \ge \frac{p_1}{p_1 - p_2} \cdot w_1 = 540$. In other words, $b \ge 540$ implies that $x_1 \ge 1$. This means that for two weight limitations b and b' with $b \ge b' \ge 540$ and $b \equiv b' \pmod{w_1}$, the optimum solutions to both problems are almost the same except that we will fulfill the part of the weight limitation $b - b'$ using the first item. Assume that we calculate $\theta(b)$ for all values of b starting with $b = 0, 1, \ldots$. Then it will be seen that $\theta(b)$ is periodic, i.e., $\theta(b) = \theta(b + 15)$ for $b \ge 26$.

We use (8.5) and (8.7) to build Tables 8.1 and 8.2 to show the periodic nature of optimum integer solutions of the knapsack problem.

Notice that in the tables, the values of $\theta(b)$ are the same for the two intervals, $b = 26$ to 40 and $b = 41$ to 55.

Next, recall that $\theta(b)$ is the difference between the optimum linear program value and the optimum integer program value. For smaller values of b, we have

$$\begin{aligned} &\theta(0) = 0 \\ &\theta(1) = (p_1 - p_5) \cdot 1 = 1.2 \\ &\theta(2) = (p_1 - p_5) \cdot 2 = 2.4 \\ &\theta(3) = (p_1 - p_5) \cdot 3 = 3.6 \\ &\theta(4) = (p_1 - p_4) \cdot 4 = 0.8 \\ &\theta(5) = (p_1 - p_4) \cdot 4 + (p_1 - p_5) \cdot 1 = 2.0 \\ &\theta(6) = (p_1 - p_4) \cdot 4 + (p_1 - p_5) \cdot 2 = 3.2 \\ &\theta(7) - (p_1 - p_3) \cdot 7 = 0.4 \\ &\theta(8) = (p_1 - p_3) \cdot 7 + (p_1 - p_5) \cdot 1 = 1.6 \\ &\text{etc.} \end{aligned}$$

If we keep calculating, we would get Table 8.1 where the $\theta(b)$ values for $b = 26$ to 40 are identical to the $\theta(b)$ values for $b = 41$ to 55.

Table 8.2 shows w_j's periodic solutions in each $b \pmod{w_1}$.

To see how Table 8.2 can be used to find $F_n(b)$ for all b, take $b = 43 \equiv 13 \pmod{15}$. In the 13th row, we have periodic solution 1, 12, which

Table 8.1

Values of b	Solutions	$\sum_j w_j x_j$	$\theta(b)$
0 ⋮ 25			
26 ⋮ 40			1.2 ⋮ 2.0
41 ⋮ 55			1.2 ⋮ 2.0

Table 8.2

b (mod w_1)	w_j's periodic solution	$\sum_{j=2}^{n} w_j x_j$	θ value
0	0	0	0
1	1	1	1.2
2	1, 1	2	2.4
3	4, 7, 7	18	3.6
4	4	4	0.8
5	1, 4	5	2.0
6	7, 7, 7	21	3.2
7	7	7	0.4
8	1, 7	8	1.6
9	12, 12	24	0.8
10	1, 12, 12	25	2.0
11	4, 7	11	1.2
12	12	12	0.4
13	1, 12	13	1.6
14	7, 7	14	0.8

means that we shall fill the knapsack with weight limitation $b = 43$ by $x_1 = 2$, $x_2 = 1$, and $x_5 = 1$. If $b = 25$, then $25 \equiv 10 \pmod{15}$. In the 10th row, we have the periodic solution 1, 12, 12 with total weight of 25. This means that x_1 should be zero if $b = 25$, or, in other words, the periodic solution does not apply. However, for $b = 40$, where $40 \equiv 10 \pmod{15}$, we have $x_1 = 1$, $x_2 = 2$, and $x_5 = 1$.

Once we have Table 8.2, we have all the optimum integer solutions for any b $(b = 0, \ldots, \infty)$:

1. Given any b, find b (mod w_1), where w_1 is the weight of the best item.
2. Find the $\theta(b)$ values for values of b until $\theta(b)$ occurs periodically.
3. Use Table 8.2 to find $\sum_{j=2}^{n} w_j x_j$.

Note that the algorithm above has a time complexity of $O(nb)$. In the next section, we shall develop another algorithm that has time complexity $O(nw_1)$. The reader could and should stop here and read the next section after he or she finishes Chapter 9, on Asymptotic Algorithms.

8.2 Landa's Algorithm

The content of this section is a simplified version of the paper published in *Research Trends in Combinational Optimization*, written by T. C. Hu, L. Landa, and M.-T. Shing.[1] This section may be read independently from the preceding section.

Mathematically, the knapsack problem is defined as follows:

$$\begin{aligned} \max \quad & v = \sum_{i=1}^{n} v_i x_i \\ \text{subject to} \quad & w = \sum_{i=1}^{n} w_i x_i \leq b \end{aligned} \tag{8.8}$$

with v_i, w_i, x_i, b all non-negative integers. The first optimal knapsack algorithm based on dynamic programming was developed by Gilmore and Gomory (1966).

Let $F_k(y)$ be the maximum value obtained in (8.8). If the knapsack has a weight-carrying capacity y and only the first k kinds of items are used, we can compute $F_k(y)$ for $k = 1, \ldots, n$ and $y = 1, \ldots, b$ as follows:

$$F_k(0) = 0 \quad \text{for all } k \text{ and}$$
$$F_1(y) = v_1 \left\lfloor \frac{y}{w_1} \right\rfloor$$

In general,

$$F_k(y) = \max\{F_{k-1}(y),\ v_k + F_k(y - w_k)\} \tag{8.9}$$

Based on (8.9), we can build a table of n rows and b columns and solve the knapsack problem in $O(nb)$ time. Note that this algorithm runs in pseudo-polynomial time because any reasonable encoding of the knapsack problem requires only a polynomial in n and $\log_2 b$.

When there are no restrictions on the values of x_i and we need to calculate the optimal solutions for large values of b, Gilmore and Gomory discovered that the optimum solutions have a periodic structure when b is sufficiently large. For simplicity, we always assume

[1] T. C. Hu, L. Landa, and M.-T. Shing, "The Unbounded Knapsack Problem," Research Trends in Combinatorial Optimization, 2009, pp. 201–217.

$$\frac{v_1}{w_1} \geq \max_j \frac{v_j}{w_j}$$

in the rest of this section, and, in the case of a tie, we let $w_1 < w_j$ so that the first item is always the best item. Note that the best item can be found in $O(n)$ time without any need to sort the items according to their value-to-weight ratios.

When the weight-carrying capacity b exceeds a critical value b^{**}, the optimal solution for b is equal to the optimal solution for $b - w_1$ plus a copy of the best item. In other words, we can first fill a portion of the knapsack using the optimal solutions for a smaller weight-carrying capacity and fill the rest of the knapsack with the best item only. Two simple upper bounds for the critical value b^{**} are

$$b^{**} \leq \sum \text{lcm}(w_1, w_j) \quad (j \neq 1) \tag{8.10}$$

and

$$b^{**} \leq \frac{w_1 p_1}{p_1 - p_2} \tag{8.11}$$

where

$p_1 = \left(\frac{v_1}{w_1}\right)$ is the highest value-to-weight ratio and
$p_2 = \left(\frac{v_2}{w_2}\right)$ is the second highest value-to-weight ratio.

For illustration purposes, we shall use the following numerical example in this section:

$$\begin{aligned} \max \quad & v = 8x_1 + 3x_2 + 17x_3 \\ \text{subject to} \quad & w = 8x_1 + 5x_2 + 18x_3 \leq b \end{aligned} \tag{8.12}$$

where b and x_i are non-negative integers.

Definition To find the starting point of the periodic optimal structure, we shall partition the weight-carrying capacity b into w_1 classes (called *threads*), where

b is in the thread 0 if $b \pmod{w_1} = 0$.
b is in the thread 1 if $b \pmod{w_1} = 1$.
...
b is in the thread $(w_1 - 1)$ if $b \pmod{w_1} = (w_1 - 1)$.

Definition We shall call the weight-carrying capacity the *boundary* and present an algorithm to pinpoint the necessary and sufficient value of the boundary in each thread (called the *thread critical boundary*). For brevity, we shall use:

$t(b)$ denotes the thread of a boundary b.
$b^*(i)$ denotes the thread critical boundary of thread i.
$b^k(i)$ denotes the thread critical boundary of thread i using the first k types of items only.

If b happens to be a multiple of w_1, then b is in the thread 0 and the critical boundary of the thread 0 is zero. For our numerical example, we have

$$\begin{array}{llll} b^*(0) = 0, & b^*(1) = 1, & b^*(2) = 18, & b^*(3) = 19 \\ b^*(4) = 36, & b^*(5) = 5, & b^*(6) = 6, & b^*(7) = 23 \end{array}$$

and

$$b^{**} = \max_j b^*(j) = b^*(4) = 36 \quad (8.13)$$

The optimal solution for any boundary b larger than the critical value $b^{**} = 36$ will exhibit the periodic structure that uses one or more copies of the best item in our example.

Landa (2004) proposes a new algorithm for computing the critical thread boundaries. Landa focuses on the *gain* from filling the otherwise unusable space (if we are limited to use integral copies of the best item only). One major advantage of using gain is that the gain computation involves only integers. And, the gain concept enables us to obtain a very simple algorithm for finding the optimal solutions when the boundary b is less than $b^*(t(b))$. To understand how Landa's algorithm works, we first formally define the concept of gain, a measure of the efficiency of different packings for a boundary b.

Definition The *gain* of a boundary b is the difference between the optimal value $V(b)$ and the value of the packing using the best item only. We define

$$g(b) = V(b) - v_1 \left\lfloor \frac{b}{w_1} \right\rfloor \quad (8.14)$$

Let $b_b = b_s + w_1$. Then b_b and b_s reside in the same thread. If the optimal solution for b_b contains at least one copy of w_1, then its sub-solution for a knapsack of the boundary b_s must also be optimal by a "principle of optimality". To formalize this concept, we state it as Lemma 8.1.

Lemma 8.1 *If the optimal solution of the boundary $b_b = b_s + w_1$ equals the optimal solution of the boundary b_s plus one copy of the best item, then the gains $g(b_b)$ and $g(b_s)$ are equal.*

Proof

$$\begin{aligned} g(b_b) &= V(b_b) - v_1 \left\lfloor \frac{b_b}{w_1} \right\rfloor = V(b_b) - v_1 \left\lfloor \frac{b_s + w_1}{w_1} \right\rfloor \\ &= V(b_s) + v_1 - v_1 \left\lfloor \frac{b_s + w_1}{w_1} \right\rfloor \\ &= V(b_s) - v_1 \left\lfloor \frac{b_s}{w_1} \right\rfloor \\ &= g(b_s) \end{aligned}$$

□

Corollary 8.1 *The gains are monotonically increasing in each thread.*

Proof Let $V(b_s)$ and $g(b_s)$ be the values of the optimal packing for b_s and its gain. We can create a feasible solution for $b_b = b_s + w_1$ by using one copy of the best item. The optimal value $V(b_b)$ must be greater than or equal to $V(b_s) + v_1$ since $b_s + w_1$ is a feasible solution. From the definition (8.14), we have $g(b_b) \geq g(b_s)$. □

From now on, if there is more than one optimal solution for a boundary b, we prefer the optimal solution with the largest x_1. Since we want to find the smallest boundary where the optimal solution becomes periodic in each thread, we shall partition boundaries in each thread into two categories.

Definition The boundary $b_b = b_s + w_1$ is a *special boundary* if its gain $g(b_b)$ is larger than $g(b_s)$, the gain of the boundary b_s. The boundary is a *non-special boundary* if $g(b_b) = g(b_s)$. Note that the definitions of special and non-special boundaries are for boundaries in the same thread.

Let us build Table 8.3 for our numerical example (8.12), where each column corresponds to a thread. Each cell has two numbers: the top number is the boundary and the bottom number is its gain. All special boundaries are marked with darker borders, and the special cell with the largest gain (i.e., thread critical boundary) in each column is marked with darker borders and shading. Since there is no cell above the first row, we define all cells in the first row as special boundaries.

Lemma 8.2 *The gain in any cell is strictly less than v_1.*

Proof By definition of gain in (8.14), we have

$$V(b) = g(b) + v_1 \left\lfloor \frac{b}{w_1} \right\rfloor$$

Table 8.3 Threaded view of optimal gains

Threads							
0	1	2	3	4	5	6	7
$b=0$ $g=0$	$b=1$ $g=0$	$b=2$ $g=0$	$b=3$ $g=0$	$b=4$ $g=0$	$b=5$ $g=3$	$b=6$ $g=3$	$b=7$ $g=3$
$b=8$ $g=0$	$b=9$ $g=0$	$b=10$ $g=0$	$b=11$ $g=0$	$b=12$ $g=0$	$b=13$ $g=3$	$b=14$ $g=3$	$b=15$ $g=3$
$b=16$ $g=0$	$b=17$ $g=0$	$b=18$ $g=1$	$b=19$ $g=1$	$b=20$ $g=1$	$b=21$ $g=3$	$b=22$ $g=3$	$b=23$ $g=4$
$b=24$ $g=0$	$b=25$ $g=0$	$b=26$ $g=1$	$b=27$ $g=1$	$b=28$ $g=1$	$b=29$ $g=3$	$b=30$ $g=3$	$b=31$ $g=4$
$b=32$ $g=0$	$b=33$ $g=0$	$b=34$ $g=1$	$b=35$ $g=1$	$b=36$ $g=2$	$b=37$ $g=3$	$b=38$ $g=3$	$b=39$ $g=4$
$b=40$ $g=0$	$b=41$ $g=0$	$b=42$ $g=1$	$b=43$ $g=1$	$b=44$ $g=2$	$b=45$ $g=3$	$b=46$ $g=3$	$b=47$ $g=4$

If $g(b) = v_1$, then

$$V(b) = v_1 + v_1\left\lfloor \frac{b}{w_1} \right\rfloor = v_1\left\lfloor \frac{b + w_1}{w_1} \right\rfloor > \frac{v_1}{w_1} \cdot b$$

a contradiction. □

Lemma 8.3 *There exists a cell in each thread that has the maximum gain, with all boundaries in the thread beyond that cell having the same gain.*

Proof We know that all gains are non-negative integers, and from Lemma 8.2, all gains are less than v_1. Since the gains are monotonically increasing in each thread by Corollary 8.1, the gain will stabilize somewhere in each thread. □

Suppose we have an optimal solution to a boundary b_s with gain $g(b_s)$, and the optimal solution for the boundary $b_b = b_s + w_k$ (in a different thread) uses the same optimal solution for b_s plus one copy of the kth item. Then the gain $g(b_b)$ of the boundary b_b is related to $g(b_s)$ based on the formula stated in Theorem 8.1.

Theorem 8.1 *If the optimal solution for the boundary $b_b = b_s + w_k$ in the thread $t(b_b)$ consists of the optimal solution for the boundary b_s in the thread $t(b_s)$ plus one copy of the kth item, then*

$$g(b_b) = g(b_s) + v_k - v_1 \left\lfloor \frac{t(b_s) + w_k}{w_1} \right\rfloor \tag{8.15}$$

Proof Since

$$g(b_b) = V(b_b) - v_1 \left\lfloor \frac{b_b}{w_1} \right\rfloor = V(b_s) + v_k - v_1 \left\lfloor \frac{b_s + w_k}{w_1} \right\rfloor \text{ and}$$

$$b_s = b_s \,(\bmod\, w_1) + w_1 \left\lfloor \frac{b_s}{w_1} \right\rfloor = t(b_s) + w_1 \left\lfloor \frac{b_s}{w_1} \right\rfloor$$

$$\begin{aligned} g(b_b) &= V(b_s) + v_k - v_1 \left(\left\lfloor \frac{t(b_s) + w_k}{w_1} \right\rfloor + \left\lfloor \frac{b_s}{w_1} \right\rfloor \right) \\ &= \left(V(b_s) - v_1 \left\lfloor \frac{b_s}{w_1} \right\rfloor \right) + v_k - v_1 \left\lfloor \frac{t(b_s) + w_k}{w_1} \right\rfloor \\ &= g(b_s) + v_k - v_1 \left\lfloor \frac{t(b_s) + w_k}{w_1} \right\rfloor \end{aligned}$$

□

Note that in Theorem 8.1, the formula (8.15) does not contain the actual value of the optimal solutions $V(b_b)$ or $V(b_s)$. It relates the gains of two cells in two different threads $t(b_b)$ and $t(b_s)$ using only the values of v_1, w_1, v_k, w_k, $t(b_s)$, and $t(b_b)$.

From now on, we describe the method for obtaining the largest gain (and its corresponding boundary) in each thread. From the principle of optimality, the sub-solution of an optimal solution must be an optimal solution of the sub-boundary. However, the principle does not say how to extend an optimal solution of a boundary to the optimal solution of a large boundary. We shall show how to use the gain formula in Theorem 8.1 to discover the gain of a larger boundary from a smaller one. For brevity, we shall use:

1. $g(b)$ to denote the gain of a boundary b
2. $g_t^*(i)$ to denote the largest gain of the thread i, which is the gain of $b^*(i)$, the critical boundary in the thread i
3. $g_t^k(i)$ to denote the gain of $b^k(i)$, the largest gain of the thread i using the first k types of items only

Starting with only one type of item with value v_1 and weight w_1, we have the thread critical boundary $b^1(i) = i$ and the gain $g_t^1(i) = 0$ for every $0 \le i \le (w_1 - 1)$.

Next, we want to pack the boundaries in each thread with the first and second types of items and see if the addition of a second type of items with value v_2 and weight w_2 can increase the maximum gain in each thread.

Definition This newly introduced type is called the *challenge* type.

Assuming that we have obtained the correct value of $g_t^2(i)$, the largest gain in thread i using the first two types of items (e.g., $g_t^2(0) = g_t^1(0) = 0$), we can use the gain formula in Theorem 8.1 to get a better gain for the thread $j = (i + w_2)(\bmod w_1)$ and set the gain $g_t^2(j)$ to

$$g_t^2(j) = \max\left\{g_t^1(j),\quad g_t^2(i) + v_2 - v_1\left\lfloor\frac{i+w_2}{w_1}\right\rfloor\right\}$$

In general, to find $g_t^k(j)$ from $g_t^k(i)$, $j = (i + w_k)(\bmod w_1)$, we set $g_t^k(j)$ to

$$g_t^k(j) = \max\left\{g_t^{k-1}(j),\quad g_t^k(i) + v_k - v_1\left\lfloor\frac{i+w_k}{w_1}\right\rfloor\right\} \tag{8.16}$$

Let us illustrate the algorithm using the numerical example (8.12).

We start with Table 8.4 for the best-item-only solution, where $b^1(i) = b^*(i) = i$ and $g_t^1(i) = 0$, with $0 \le i \le 7$.

Table 8.4 Thread critical boundaries using the first item only

	Threads							
K	0	1	2	3	4	5	6	7
1	$b=0$ $g=0$	$b=1$ $g=0$	$b=2$ $g=0$	$b=3$ $g=0$	$b=4$ $g=0$	$b=5$ $g=0$	$b=6$ $g=0$	$b=7$ $g=0$

Next, we introduce the second item with $v_2 = 3$ and $w_2 = 5$. We start with thread 0 and try to improve the gain of thread 5 using one copy of the second item.

Since $0 + 3 - 8\left\lfloor\frac{0+5}{8}\right\rfloor = 3 > 0$, we set $b^2(5) = 5$ and $g_t^2(5) = 3$. We then look at $b = 10$ in thread 2. We have $3 + 3 - 8\left\lfloor\frac{5+5}{8}\right\rfloor = -2 < 0$; we set $b^2(2) = b^1(2) = 2$ and $g_t^2(2) = g_t^1(2) = 0$. From thread 2, we will visit the remaining threads in the order of 7, 4, 1, 6, 3 and use the formula (8.16) to obtain the thread critical boundaries and the gains shown in Table 8.5.

Table 8.5 Thread critical boundaries using the item types 1 and 2

	Threads							
K	0	1	2	3	4	5	6	7
2	$b=0$ $g=0$	$b=1$ $g=0$	$b=2$ $g=0$	$b=3$ $g=0$	$b=4$ $g=0$	$b=5$ $g=3$	$b=6$ $g=3$	$b=7$ $g=3$

Table 8.5 thus contains the thread critical boundaries using the two types of items.

Definition The sequence of threads visited (in this case, we have 0, 5, 2, 7, 4, 1, 6, 3) is called a *chain*.

Because $\gcd(w_1, w_2) = \gcd(8, 5) = 1$, we are able to visit all threads in one chain. Now, let us introduce the third item with $v_3 = 17$ and $w_3 = 18$.

Since $\gcd(w_1, w_3) = \gcd(8, 18) = 2 \neq 1$, we cannot visit all the threads. We can only visit 0, 2, 4, 6 in one chain if we start with thread 0. Instead of visiting all the threads in a single chain, we need two chains. The second chain involves the threads 1, 3, 5, and 7.

Definition The technique of finding the thread which does not use the newly introduced type is called the *double-loop technique*, which will be used whenever $\gcd(w_1, w_k) \neq 1$.

There exists at least one cell in the chain which does not use any k^{th}-type item. Hence, for our numerical example, there must exist a thread $j \in \{1, 3, 5, 7\}$ such that

$$g_t^3(j) = g_t^2(j) \geq g_t^3(i) + 17 - 8\left\lfloor \frac{i+18}{8} \right\rfloor \geq g_t^2(i) + 17 - 8\left\lfloor \frac{i+18}{w_1} \right\rfloor$$

Table 8.6 shows the values of $g_t^3(j)$ and $b^3(j)$ of the threads 7, 1, 3, and 5 after the first loop. Note that both threads 1 and 5 can serve as the starting cell for the chain.

Table 8.6 The $g_t^3(j)$ and $b^3(j)$ values after the first loop

i	j	$g_t^2(j)$	$g_t^3(i) + 17 - 8 \times \left\lfloor \frac{i+18}{8} \right\rfloor$	$g_t^3(j)$	$b^3(j)$	Remarks
5	7	3	4	4	23	
7	1	0	−3	0	1	Possible starting cell for the chain 1, 3, 5, 7
1	3	0	1	1	19	
3	5	3	3	3	5	Possible starting cell for the chain 1, 3, 5, 7

Then we start with the thread 5 and visit threads in the second chain in the order of 7, 1, and 3 one more time, to obtain the thread critical boundaries and gains shown in Table 8.7.

Table 8.7 Thread critical boundaries using the item types 1, 2, and 3

	Threads							
K	0	1	2	3	4	5	6	7
3	$b = 0$ $g = 0$	$b = 1$ $g = 0$	$b = 18$ $g = 1$	$b = 19$ $g = 1$	$b = 36$ $g = 2$	$b = 5$ $g = 3$	$b = 6$ $g = 3$	$b = 23$ $g = 4$

Hence, we again have

$$\begin{aligned} &b^*(0) = 0, \quad b^*(1) = 1, \quad b^*(2) = 18, \quad b^*(3) = 19 \\ &b^*(4) = 36, \quad b^*(5) = 5, \quad b^*(6) = 6, \quad b^*(7) = 23 \end{aligned}$$

and

$$b^{**} = \max_j b^*(j) = b^*(4) = 36$$

High-Level Description of Landa's Algorithm

The algorithm for finding the thread critical boundaries:

1. Find the best item type, i.e., the item type with the highest value-to-weight ratio, and name it as the first item. In case of a tie, pick the item with the smallest weight.
2. Create an array T with w_1 entries. Initialize the entry $T[i]$, $0 \leq i \leq w_1 - 1$, with the ordered pair $\left(b^1(i) = i,\ g_t^1(i) = 0\right)$.
3. For $k = 2, \ldots, n$, introduce the challenge types, one type at a time. Starting from the thread 0, we set $b^k(0) = 0$ and $g_t^k(0) = 0$, traverse the threads in the first chain in the order of $j = w_k \,(\mathrm{mod}\, w_1), 2w_k \,(\mathrm{mod}\, w_1), \ldots$, compute the $(b^*(j), g_t^k(j))$ using the formula (8.16), and update entries in T accordingly. If $\gcd(w_1, w_k) > 1$, use the double-loop technique to compute $(b^k(j), g_t^k(j))$ for the threads in each of the remaining chains and update the entries in the array T accordingly.

Since Step 1 takes $O(n)$ time, Step 2 takes $O(w_1)$ time, and Step 3 takes $O(w_1)$ time for each new item type and there are $n - 1$ new item types, Landa's algorithm runs in $O(nw_1)$ time.

8.3 Exercises

1. What is the simplest integer program?
2. If an integer program has an optimal value x_0, what is a non-integer least upper bound? If the integer program has a minimal value z, what is the greatest non-integer lower bound?
3. If the given data of an integer program are all multiples of ten, including the right-hand side, and we just need the optimum solution of the given integer program to be in integers, would this make the integer program easier to solve? Why or why not?
4. If we do not get integer solutions, what are the reasons for this?

9 Asymptotic Algorithms

9.1 Solving Integer Programs

Consider an integer program

$$\begin{aligned} \min \quad & z = x_1 + x_2 + x_3 + x_4 \\ \text{subject to} \quad & \begin{bmatrix} 3 \\ 1 \end{bmatrix} x_1 + \begin{bmatrix} 1 \\ 3 \end{bmatrix} x_2 + \begin{bmatrix} 2 \\ 1 \end{bmatrix} x_3 + \begin{bmatrix} 1 \\ 2 \end{bmatrix} x_4 = \begin{bmatrix} 11 \\ 11 \end{bmatrix} \\ & x_j \geq 0 \quad \text{integers} \end{aligned} \tag{9.1}$$

Definition If we drop the integer requirements on the variables of an integer program, then the resulting linear program is called the *associated linear program.*

So, in the case of (9.1), the optimum solution of the associated linear program is $x_1 = \frac{11}{4}, x_2 = \frac{11}{4}$, and $x_3 = x_4 = 0$. In other words, x_1 and x_2 are the basic variables, and x_3 and x_4 are the non-basic variables.

In Figure 9.1, the black dots shown are the coordinates that could be obtained as sums of integer multiples of the two basic vectors $\begin{bmatrix} 3 \\ 1 \end{bmatrix}$ and $\begin{bmatrix} 1 \\ 3 \end{bmatrix}$.

T.C. Hu, A.B. Kahng, *Linear and Integer Programming Made Easy*,
DOI 10.1007/978-3-319-24001-5_9

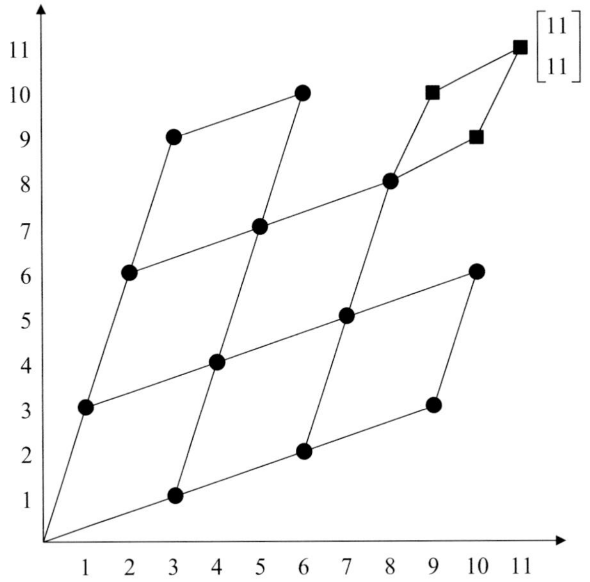

Fig. 9.1 Possible coordinates of sums of integer multiples of the vectors $\begin{bmatrix} 3 \\ 1 \end{bmatrix}$ and $\begin{bmatrix} 1 \\ 3 \end{bmatrix}$ from (9.1)

Since the solutions of the associated linear program are not integers, the only way to solve it is to increase the values of the non-basic variables x_3 and x_4.

In solving the associated linear program, we multiply by the inverse of the matrix

$$\begin{bmatrix} 3 & 1 \\ 1 & 3 \end{bmatrix} \quad \text{with} \quad \begin{bmatrix} 3 & 1 \\ 1 & 3 \end{bmatrix}^{-1} = \frac{1}{8}\begin{bmatrix} 3 & -1 \\ -1 & 3 \end{bmatrix}$$

The result of multiplying $\frac{1}{8}\begin{bmatrix} 3 & -1 \\ -1 & 3 \end{bmatrix}$ by the constraint of (9.1) is shown as (9.2). The constraint of (9.1) becomes a congruence relation, where all numbers are equal (mod 1).

$$\begin{bmatrix} 1 \\ 0 \end{bmatrix} x_1 + \begin{bmatrix} 0 \\ 1 \end{bmatrix} x_2 + \begin{bmatrix} \frac{5}{8} \\ \frac{1}{8} \end{bmatrix} x_3 + \begin{bmatrix} \frac{1}{8} \\ \frac{5}{8} \end{bmatrix} x_4 = \begin{bmatrix} \frac{6}{8} \\ \frac{6}{8} \end{bmatrix} \tag{9.2}$$

To satisfy the relation (9.2), we see by inspection that $x_3 = 1$ and $x_4 = 1$. Then, we obtain the optimum integer solution by substituting x_3 and x_4 into (9.1) and solving the reduced integer program

$$\begin{aligned} \min \quad & z = x_1 + x_2 + x_3 + x_4 \\ \text{subject to} \quad & \begin{bmatrix} 3 \\ 1 \end{bmatrix} x_1 + \begin{bmatrix} 1 \\ 3 \end{bmatrix} x_2 + \begin{bmatrix} 2 \\ 1 \end{bmatrix} x_3 + \begin{bmatrix} 1 \\ 2 \end{bmatrix} x_4 = \begin{bmatrix} 8 \\ 8 \end{bmatrix} \\ & x_j \geq 0 \quad \text{integers} \end{aligned} \tag{9.3}$$

The solution is $x_1 = 2, x_2 = 2, x_3 = 0$, and $x_4 = 0$ with $z = 4$. Note that the RHS of (9.3) is from the RHS of (9.1),

$$\begin{bmatrix} 11 \\ 11 \end{bmatrix} - \begin{bmatrix} 2 \\ 1 \end{bmatrix} - \begin{bmatrix} 1 \\ 2 \end{bmatrix} = \begin{bmatrix} 8 \\ 8 \end{bmatrix}$$

In this example, we first solve the associated linear program. If basic variables are not integers, we solve the equivalent constraint relation by increasing the values of non-basic variables.

Let us use matrix notation and describe the general approach.

Consider an integer program

$$\begin{aligned} \max \quad & z = \boldsymbol{c}'\boldsymbol{x}' \\ \text{subject to} \quad & \boldsymbol{A}'\boldsymbol{x}' \leq \boldsymbol{b} \\ & \boldsymbol{x}' \geq \boldsymbol{0} \quad \text{integers} \end{aligned} \tag{9.4}$$

where $\boldsymbol{A}'$ is an $m \times n$ integer matrix, $\boldsymbol{b}$ is an integer m-vector, and $\boldsymbol{c}'$ is an integer n-vector.

Alternatively, the integer program (9.4) can be written as

$$\begin{aligned} \max \quad & z = \boldsymbol{c}\boldsymbol{x} \\ \text{subject to} \quad & \boldsymbol{A}\boldsymbol{x} = \boldsymbol{b} \\ & \boldsymbol{x} \geq \boldsymbol{0} \quad \text{integers} \end{aligned} \tag{9.5}$$

where $\boldsymbol{A}$ is an $m \times (m+n)$ integer matrix, $\boldsymbol{c}$ is an integer $(m+n)$-vector, and $\boldsymbol{x}$ is an integer $(m+n)$-vector which includes the slack variables introduced to convert the inequalities of (9.4) to the equalities of (9.5). For simplicity, we shall assume that $\boldsymbol{A}$ contains an $m \times m$ identity matrix $\boldsymbol{I}$. Partitioning $\boldsymbol{A}$ as $[\boldsymbol{B}, \boldsymbol{N}]$, we can write (9.5) as

$$\begin{aligned} \max \quad & z = \boldsymbol{c}_B\boldsymbol{x}_B + \boldsymbol{c}_N\boldsymbol{x}_N \\ \text{subject to} \quad & \boldsymbol{B}\boldsymbol{x}_B + \boldsymbol{N}\boldsymbol{x}_N = \boldsymbol{b} \\ & \boldsymbol{x}_B, \boldsymbol{x}_N \geq \boldsymbol{0} \quad \text{integers} \end{aligned} \tag{9.6}$$

where $\boldsymbol{B}$ is an $m \times n$ nonsingular matrix. Expressing $\boldsymbol{x}_B$ in terms of $\boldsymbol{x}_N$, i.e., $\boldsymbol{x}_B = \boldsymbol{B}^{-1}\boldsymbol{b} - \boldsymbol{B}^{-1}\boldsymbol{N}\boldsymbol{x}_N$, (9.6) becomes

$$\begin{aligned} \max \quad & z = \boldsymbol{c}_B \boldsymbol{B}^{-1} \boldsymbol{b} - \left(\boldsymbol{c}_B \boldsymbol{B}^{-1} \boldsymbol{N} - \boldsymbol{c}_N \right) \boldsymbol{x}_N \\ \text{subject to} \quad & \boldsymbol{x}_B + \boldsymbol{B}^{-1} \boldsymbol{N} \boldsymbol{x}_N = \boldsymbol{B}^{-1} \boldsymbol{b} \\ & \boldsymbol{x}_B, \boldsymbol{x}_N \geq \boldsymbol{0} \quad \text{integers} \end{aligned} \tag{9.7}$$

If we consider (9.7) as a linear program, i.e., drop the integer restriction on $\boldsymbol{x}_B$ and $\boldsymbol{x}_N$, and if $\boldsymbol{B}$ is the optimum basis of the linear program, then the optimum solution to the linear program is

$$\boldsymbol{x}_B = \boldsymbol{B}^{-1}\boldsymbol{b}, \quad \boldsymbol{x}_N = \boldsymbol{0}$$

where

$$\boldsymbol{c}_B \boldsymbol{B}^{-1} \boldsymbol{N} - \boldsymbol{c}_N \geq \boldsymbol{0}$$

If $\boldsymbol{B}^{-1}\boldsymbol{b}$ happens to be an integer vector, then $\boldsymbol{x}_B = \boldsymbol{B}^{-1}\boldsymbol{b}, \boldsymbol{x}_N = \boldsymbol{0}$ is also the optimum solution to the integer program (9.7). If $\boldsymbol{B}^{-1}\boldsymbol{b}$ is not an integer vector, then we must increase $\boldsymbol{x}_N$ from zero to some non-negative integer vector such that

$$\boldsymbol{x}_B = \boldsymbol{B}^{-1}\boldsymbol{b} - \boldsymbol{B}^{-1}\boldsymbol{N}\boldsymbol{x}_N \geq \boldsymbol{0}, \quad \text{integers} \tag{9.8}$$

Consider the numerical example

$$\begin{aligned} \max \quad & z = -4x_3 - 5x_4 \\ \text{subject to} \quad & -3x_3 - x_4 + x_5 = -2 \\ & -x_3 - 4x_4 + x_1 = -5 \\ & -3x_3 - 2x_4 + x_2 = -7 \\ & x_j \geq 0 \quad \text{integers} \quad (j = 1, \ldots, 5) \end{aligned}$$

The associated linear program has the optimum basis

$$\boldsymbol{B} = \begin{bmatrix} -3 & -1 & 1 \\ -1 & -4 & 0 \\ -3 & -2 & 0 \end{bmatrix}$$

with $|\det \boldsymbol{B}| = 10$. Applying

$$\boldsymbol{B}^{-1} = \begin{bmatrix} 0 & \frac{2}{10} & -\frac{4}{10} \\ 0 & -\frac{3}{10} & \frac{1}{10} \\ 1 & \frac{3}{10} & -\frac{11}{10} \end{bmatrix}$$

to the constraints in our numerical example, we have

$$\max \quad z = -\left(\frac{7}{10}\right)x_1 - \left(\frac{11}{10}\right)x_2 - \frac{112}{10}$$

and the congruence relation becomes

$$\begin{bmatrix}1\\0\\0\end{bmatrix}x_3 + \begin{bmatrix}0\\1\\0\end{bmatrix}x_4 + \begin{bmatrix}0\\0\\1\end{bmatrix}x_5 + \begin{bmatrix}\frac{2}{10}\\-\frac{3}{10}\\\frac{3}{10}\end{bmatrix}x_1 + \begin{bmatrix}-\frac{4}{10}\\\frac{1}{10}\\-\frac{11}{10}\end{bmatrix}x_2 = \begin{bmatrix}\frac{18}{10}\\\frac{8}{10}\\\frac{42}{10}\end{bmatrix}.$$

Therefore, we want

$$\begin{aligned}\min \quad & \left(\frac{7}{10}\right)x_1 + \left(\frac{11}{10}\right)x_2 \\ \text{subject to} \quad & \begin{bmatrix}\frac{2}{10}\\\frac{7}{10}\\\frac{3}{10}\end{bmatrix}x_1 + \begin{bmatrix}\frac{6}{10}\\\frac{1}{10}\\\frac{9}{10}\end{bmatrix}x_2 \equiv \begin{bmatrix}\frac{8}{10}\\\frac{8}{10}\\\frac{2}{10}\end{bmatrix}\end{aligned} \tag{9.9}$$

By inspection, we have $x_1 = 1$ and $x_2 = 1$ as the optimum solution of (9.9).

In summary, we solve an integer program in three steps:

1. Treat the integer program as a linear program. If the associated linear program has an optimum solution in integers, then the optimum solution is also the optimum solution of the original integer program.
2. If not, map all non-basic columns of the associated linear program and its RHS into group elements and find the cheapest way of expressing the RHS.
3. Substitute the values of the non-basic variables from Step 2 into the original integer program, and obtain the values of basic variables.

It turns out that we cannot solve all integer programs simply by following these three steps. However, most integer programs with large RHS $\boldsymbol{b}$ can be solved with the above three steps.

Let us consider another integer program as shown in (9.10).

$$\begin{aligned}\min \quad & z = x_1 + x_2 + x_3 + x_4 \\ \text{subject to} \quad & \begin{bmatrix}3\\2\end{bmatrix}x_1 + \begin{bmatrix}2\\3\end{bmatrix}x_2 + \begin{bmatrix}1\\2\end{bmatrix}x_3 + \begin{bmatrix}1\\5\end{bmatrix}x_4 = \begin{bmatrix}11\\11\end{bmatrix} \\ & x_j \geq 0 \quad \text{integers}\end{aligned} \tag{9.10}$$

This is shown in Figure 9.2. There are three straight lines from the origin: one straight line with multiples of $\begin{bmatrix} 3 \\ 2 \end{bmatrix}$, another straight line with multiples of $\begin{bmatrix} 2 \\ 3 \end{bmatrix}$, and the third straight line with multiples of $\begin{bmatrix} 1 \\ 2 \end{bmatrix}$. The point $\begin{bmatrix} 11 \\ 11 \end{bmatrix}$ is marked with a black square. The point with coordinates $\begin{bmatrix} 8 \\ 13 \end{bmatrix}$ is also shown as a black square (to be used in Example 2, below).

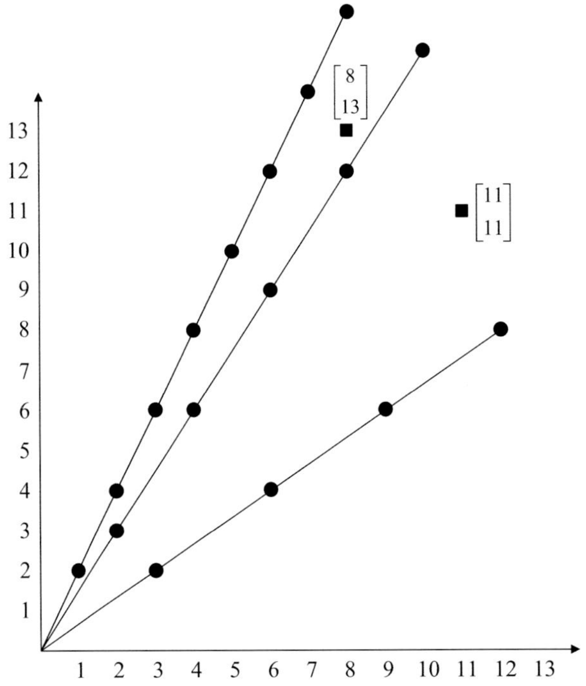

Fig. 9.2 Possible coordinates of integer multiples of the integer program as shown in (9.10)

Note that the line from the origin to the point $\begin{bmatrix} 11 \\ 11 \end{bmatrix}$ is in the cone spanned by $\begin{bmatrix} 3 \\ 2 \end{bmatrix}$ and $\begin{bmatrix} 2 \\ 3 \end{bmatrix}$. The associated linear program of (9.10) has its basic variables x_1 and x_2 with their values

$$x_1 = \frac{11}{5}, \quad x_2 = \frac{11}{5}, \quad x_3 = 0, \quad x_4 = 0, \quad \text{and} \quad z = \frac{22}{5}$$

since $\begin{bmatrix} 3 & 2 \\ 2 & 3 \end{bmatrix}^{-1} = \frac{1}{5}\begin{bmatrix} 3 & -2 \\ -2 & 3 \end{bmatrix}$.

The congruence relation is obtained by multiplying by the inverse of the matrix $\begin{bmatrix} 3 & 2 \\ 2 & 3 \end{bmatrix}$, i.e.,

$$\frac{1}{5}\begin{bmatrix} 3 & -2 \\ -2 & 3 \end{bmatrix}\begin{bmatrix} 3 & 2 & 1 & 1 \\ 2 & 3 & 2 & 5 \end{bmatrix} = \frac{1}{5}\begin{bmatrix} 5 & 0 & -1 & -7 \\ 0 & 5 & 4 & 13 \end{bmatrix}$$

$$\frac{1}{5}\begin{bmatrix} 3 & -2 \\ -2 & 3 \end{bmatrix}\begin{bmatrix} 11 \\ 11 \end{bmatrix} = \frac{1}{5}\begin{bmatrix} 11 \\ 11 \end{bmatrix}$$

The result is

$$\begin{bmatrix} 1 & 0 & -\frac{1}{5} & -\frac{7}{5} & \frac{11}{5} \\ 0 & 1 & \frac{4}{5} & \frac{13}{5} & \frac{11}{5} \end{bmatrix}$$

$$\text{or} \quad \begin{bmatrix} \frac{4}{5} \\ \frac{4}{5} \end{bmatrix} x_3 + \begin{bmatrix} \frac{3}{5} \\ \frac{3}{5} \end{bmatrix} x_4 \equiv \begin{bmatrix} \frac{1}{5} \\ \frac{1}{5} \end{bmatrix} \tag{9.11}$$

with many possible solutions to (9.11):

$$\begin{cases} x_3 = 4, & x_4 = 0 \\ x_3 = 0, & x_4 = 2 \\ x_3 = 2, & x_4 = 1 \end{cases}$$

These will force x_1 or x_2 to be negative.

Example 1

$$\begin{aligned} \min \quad & z = x_1 + x_2 + x_3 + x_4 \\ \text{subject to} \quad & \begin{bmatrix} 3 \\ 2 \end{bmatrix} x_1 + \begin{bmatrix} 2 \\ 3 \end{bmatrix} x_2 + \begin{bmatrix} 1 \\ 2 \end{bmatrix} x_3 + \begin{bmatrix} 1 \\ 5 \end{bmatrix} x_4 = \begin{bmatrix} 44 \\ 44 \end{bmatrix} \\ & x_j \pmod 1 = 0 \text{ and } x_j \geq 0 \quad (j = 1, 2, 3, 4) \end{aligned} \tag{9.12}$$

The associated linear program has the optimum solution

$$x_1 = \frac{44}{5}, \quad x_2 = \frac{44}{5}, \quad x_3 = 0, \quad x_4 = 0, \quad \text{and} \quad z = \frac{88}{5}.$$

All the non-basic columns and the RHS are mapped into the congruence relation

$$\frac{1}{5}\begin{bmatrix} -1 \\ 4 \end{bmatrix} x_3 + \frac{1}{5}\begin{bmatrix} -7 \\ 13 \end{bmatrix} x_4 = \frac{1}{5}\begin{bmatrix} 44 \\ 44 \end{bmatrix}$$
$$\text{or} \quad \begin{bmatrix} \frac{4}{5} \\ \frac{4}{5} \end{bmatrix} x_3 + \begin{bmatrix} \frac{3}{5} \\ \frac{3}{5} \end{bmatrix} x_4 \equiv \begin{bmatrix} \frac{4}{5} \\ \frac{4}{5} \end{bmatrix} \tag{9.13}$$

where the cheapest solution is $x_3 = 1$ and $x_4 = 0$. When we substitute the values of $x_3 = 1$ and $x_4 = 0$ back to (9.12), we get the optimum integer solution of (9.12):

$$x_1 = 9, \quad x_2 = 8, \quad x_3 = 1, \quad \text{and} \quad x_4 = 0 \quad \text{with} \quad z = 18 > \frac{88}{5} = 17.6.$$

Here we are very lucky. We pay very little to make all variables to be integers.

Let us try a different RHS $\boldsymbol{b}$ as shown in (9.12a), which is the same as (9.10):

$$\begin{aligned} \min \quad & z = x_1 + x_2 + x_3 + x_4 \\ \text{subject to} \quad & \begin{bmatrix} 3 \\ 2 \end{bmatrix} x_1 + \begin{bmatrix} 2 \\ 3 \end{bmatrix} x_2 + \begin{bmatrix} 1 \\ 2 \end{bmatrix} x_3 + \begin{bmatrix} 1 \\ 5 \end{bmatrix} x_4 = \begin{bmatrix} 11 \\ 11 \end{bmatrix} \\ & x_j \geq 0 \quad \text{integers} \end{aligned} \tag{9.12a}$$

The associated linear program has the optimum solution

$$x_1 = \frac{11}{5}, \quad x_2 = \frac{11}{5}, \quad x_3 = 0, \quad x_4 = 0, \quad \text{and} \quad z = \frac{22}{5}.$$

And the congruence relation

$$\begin{bmatrix} \frac{-1}{5} \\ \frac{4}{5} \end{bmatrix} x_3 + \begin{bmatrix} \frac{-7}{5} \\ \frac{3}{5} \end{bmatrix} x_4 = \begin{bmatrix} \frac{11}{5} \\ \frac{11}{5} \end{bmatrix}$$
$$\text{or} \quad \begin{bmatrix} \frac{4}{5} \\ \frac{4}{5} \end{bmatrix} x_3 + \begin{bmatrix} \frac{3}{5} \\ \frac{3}{5} \end{bmatrix} x_4 \equiv \begin{bmatrix} \frac{1}{5} \\ \frac{1}{5} \end{bmatrix} \tag{9.13a}$$

(1) $x_3 = 0$ and $x_4 = 2$ with $x_1 = 5$ and $x_2 = -3$.
(2) $x_3 = 4$ and $x_4 = 0$ with $x_1 = 3$ and $x_2 = -1$.

Both (1) and (2) are not acceptable since $x_2 < 0$. We can solve the problem nicely with the RHS $\begin{bmatrix} 44 \\ 44 \end{bmatrix}$, but not with $\begin{bmatrix} 11 \\ 11 \end{bmatrix}$. Why? We shall explain below.

Let us now return to Example 1:

$$\begin{array}{ll} \min & z = x_1 + x_2 + x_3 + x_4 \\ \text{subject to} & \begin{bmatrix} 3 \\ 2 \end{bmatrix} x_1 + \begin{bmatrix} 2 \\ 3 \end{bmatrix} x_2 + \begin{bmatrix} 1 \\ 2 \end{bmatrix} x_3 + \begin{bmatrix} 1 \\ 5 \end{bmatrix} x_4 = \begin{bmatrix} 44 \\ 44 \end{bmatrix} \\ \text{all} & x_j (\text{mod } 1) = 0 \text{ and } x_j \geq 0 \end{array} \tag{9.12b}$$

As noted above, the associated linear program has the optimum solution

$$x_1 = \frac{44}{5}, \quad x_2 = \frac{44}{5}, \quad x_3 = 0, \quad x_4 = 0, \quad \text{and} \quad z = \frac{88}{5}.$$

And the congruence relation becomes

$$\begin{bmatrix} \frac{4}{5} \\ \frac{4}{5} \end{bmatrix} x_3 + \begin{bmatrix} \frac{3}{5} \\ \frac{3}{5} \end{bmatrix} x_4 \equiv \begin{bmatrix} \frac{4}{5} \\ \frac{4}{5} \end{bmatrix} \tag{9.13b}$$

with the cheapest solution $x_3 = 1$ and $x_4 = 0$.

When we substitute the values of $x_3 = 1$ and $x_4 = 0$ back into (9.12b), we obtain the optimum integer solution

$$x_1 = 9, \quad x_2 = 8, \quad x_3 = 1, \quad \text{and} \quad x_4 = 0.$$

When we substitute the values of $x_3 = 1$ and $x_4 = 0$ back into (9.12a), we cannot obtain the feasible integer solution that satisfies $x_j \geq 0$ for all j.

The reason that the RHS $\begin{bmatrix} 44 \\ 44 \end{bmatrix}$ works, but the RHS $\begin{bmatrix} 11 \\ 11 \end{bmatrix}$ does not, is that $\begin{bmatrix} 44 \\ 44 \end{bmatrix}$ is in between the two generators (i.e., $\begin{bmatrix} 3 \\ 2 \end{bmatrix}$ and $\begin{bmatrix} 2 \\ 3 \end{bmatrix}$) with more room to move. This illustrates how we can solve an integer program more easily if we have a larger RHS $\boldsymbol{b}$.

Example 2

$$\begin{array}{ll} \min & z = x_1 + x_2 + x_3 + x_4 \\ \text{subject to} & \begin{bmatrix} 3 \\ 2 \end{bmatrix} x_1 + \begin{bmatrix} 2 \\ 3 \end{bmatrix} x_2 + \begin{bmatrix} 1 \\ 2 \end{bmatrix} x_3 + \begin{bmatrix} 1 \\ 5 \end{bmatrix} x_4 = \begin{bmatrix} 8 \\ 13 \end{bmatrix} \\ & x_j \geq 0 \quad \text{integers} \end{array} \tag{9.14}$$

Note that $\begin{bmatrix} 8 \\ 13 \end{bmatrix}$ is spanned by $\begin{bmatrix} 3 \\ 2 \end{bmatrix}$ and $\begin{bmatrix} 1 \\ 2 \end{bmatrix}$ and is also spanned by $\begin{bmatrix} 2 \\ 3 \end{bmatrix}$ and $\begin{bmatrix} 1 \\ 2 \end{bmatrix}$. Which pair should we choose?

If we choose $\begin{bmatrix} 2 \\ 3 \end{bmatrix}$ and $\begin{bmatrix} 1 \\ 2 \end{bmatrix}$, we obtain the optimum associated linear program solution

$$x_2 = 3, \quad x_3 = 2, \quad \text{and} \quad z = 5.$$

If we choose $\begin{bmatrix} 3 \\ 2 \end{bmatrix}$ and $\begin{bmatrix} 1 \\ 2 \end{bmatrix}$, we obtain the solution

$$x_1 = 0.75, \quad x_3 = 5.75, \quad \text{and} \quad z = 6.5.$$

When we choose $\begin{bmatrix} 2 \\ 3 \end{bmatrix}$ and $\begin{bmatrix} 1 \\ 2 \end{bmatrix}$, we again have more room to move in the cone to find the optimum integer solution. Many integer problems with large RHS $\boldsymbol{b}$ can be solved with this approach.

9.2 Exercises

1. What is a congruence relation? Is the number $-\frac{1}{4}$ congruent to $\frac{1}{4}$?
2. What is the associated linear program of an integer program?
3. What does it mean for two numbers to be relatively prime?

The World Map of Integer Programs 10

Integer programs are very hard to solve. Even the knapsack problem, one of the simplest integer programs, is NP-complete. In order to solve this problem, most people would use the dynamic programming-type algorithm of Gilmore and Gomory. This algorithm is pseudo-polynomial and has time complexity $O(nb)$, where b is the weight-carrying capacity of the knapsack. The knapsack problem can also be solved using Landa's algorithm, as we saw in Section 8.2. Landa's algorithm is also pseudo-polynomial and has time complexity $O(nw_1)$, where w_1 is the weight of the best item.

When there are two constraints in an integer program, i.e., the RHS $\boldsymbol{b} = [b_1, b_2]$, we can see many nice features of integer programs. We illustrate these features in Figure 10.1. In this figure, there are two cones: the outer cone and the inner cone. The outer cone, with its origin at $\begin{bmatrix} 0 \\ 0 \end{bmatrix}$, is composed of two thick lines that are extended by the vectors $\begin{bmatrix} 2 \\ 1 \end{bmatrix}$ and $\begin{bmatrix} 1 \\ 3 \end{bmatrix}$. The inner cone, with its origin at $\begin{bmatrix} 3 \\ 4 \end{bmatrix}$, is composed of two thin lines that are parallel to the two thick lines.

We solve an integer program

$$\begin{aligned} \max \quad & x_0 = c_1x_1 + c_2x_2 + \cdots \\ \text{subject to} \quad & \begin{bmatrix} 2 \\ 1 \end{bmatrix} x_1 + \begin{bmatrix} 1 \\ 3 \end{bmatrix} x_2 + \cdots = \begin{bmatrix} b_1 \\ b_2 \end{bmatrix} \\ & x_j \geq 0 \text{ integers} \end{aligned} \tag{10.1}$$

If the associated linear program has $\begin{bmatrix} 2 & 1 \\ 1 & 3 \end{bmatrix}$ as its optimal basis, then the integer program (10.1) also has the same optimal basis provided that $\begin{bmatrix} b_1 \\ b_2 \end{bmatrix}$ can be expressed as integer combinations of $\begin{bmatrix} 2 \\ 1 \end{bmatrix}$ and $\begin{bmatrix} 1 \\ 3 \end{bmatrix}$ and are images of some non-basic columns.

T.C. Hu, A.B. Kahng, *Linear and Integer Programming Made Easy*,
DOI 10.1007/978-3-319-24001-5_10

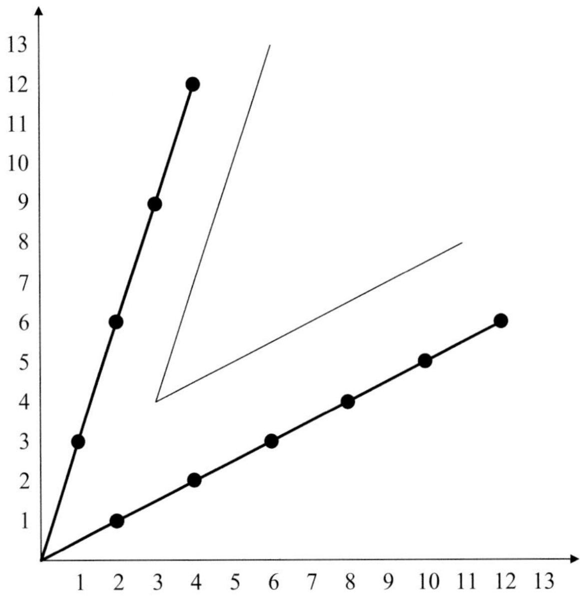

Fig. 10.1 The outer cone and inner cone of the integer program in (10.1) corresponding to basis vectors $\begin{bmatrix} 2 & 1 \\ 1 & 3 \end{bmatrix}$

If the congruence relation cannot be solved or if one of the basic variables, x_1 or x_2, is forced to be negative, then we find another basis of the associated linear program (e.g., the second-best basis of the linear program) which could be as shown in Figure 10.2. There are again two cones in Figure 10.2, where the outer cone with the two thick lines is extended by the vectors $\begin{bmatrix} 4 \\ 3 \end{bmatrix}$ and $\begin{bmatrix} 1 \\ 5 \end{bmatrix}$ and the inner cone with its origin at $\begin{bmatrix} 5 \\ 8 \end{bmatrix}$ is composed of two thin lines:

$$\begin{aligned} \max \quad & x_0 = c_1x_1 + c_2x_2 + c_3x_3 + c_4x_4 + \cdots \\ \text{subject to} \quad & \begin{bmatrix} 2 \\ 1 \end{bmatrix} x_1 + \begin{bmatrix} 1 \\ 3 \end{bmatrix} x_2 + \begin{bmatrix} 4 \\ 3 \end{bmatrix} x_3 + \begin{bmatrix} 1 \\ 5 \end{bmatrix} x_4 + \cdots = \begin{bmatrix} b_1 \\ b_2 \end{bmatrix} \\ & x_j \geq 0 \quad \text{integers} \end{aligned} \tag{10.2}$$

In summary, there are four regions in Figure 10.1 and in Figure 10.2: (*i*), (*iia*), (*iib*), and (*iii*):

(*i*) The region inside the inner cone, with the RHS $[b_1, b_2]$ very far away from $\begin{bmatrix} 3 \\ 4 \end{bmatrix}$ in Figure 10.1 or from $\begin{bmatrix} 5 \\ 8 \end{bmatrix}$ in Figure 10.2. If $[b_1, b_2]$ is in this region,

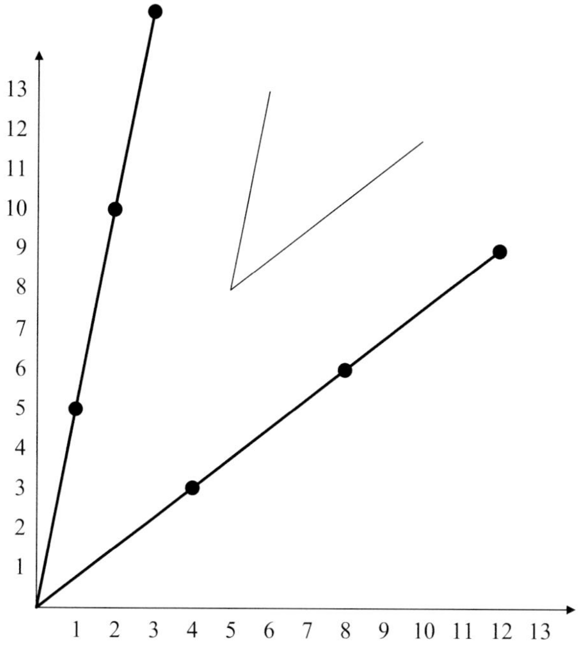

Fig. 10.2 The outer cone and inner cone of the integer program in (10.2) corresponding to basis vectors $\begin{bmatrix} 4 & 1 \\ 3 & 5 \end{bmatrix}$

(1) we have to solve the congruence relation, and (2) we do not have to worry that the basic variables of the associated linear program would be forced to be negative.

(*ii*) The $[b_1, b_2]$ lies within the strips that are between parallel thin and thick lines in the figure. We use (*iia*) to denote the upper strip and (*iib*) to denote the lower strip. These strips extend farther and farther away from the origin. (Note that the two strips could be of different widths and vary from problem to problem.) In this case, (1) we have to solve the congruence relation, and (2) we should consider that the basic variables of the associated linear program would be forced to be negative. If the $[b_1, b_2]$ values are very large, the final integer solution may contain the vectors $\begin{bmatrix} 2 \\ 1 \end{bmatrix}$ and $\begin{bmatrix} 1 \\ 3 \end{bmatrix}$ many times over as in Figure 10.1 or the vectors $\begin{bmatrix} 4 \\ 3 \end{bmatrix}$ and $\begin{bmatrix} 1 \\ 5 \end{bmatrix}$ many times over as in Figure 10.2.

(*iii*) This region is the intersection of the two strips of constant widths corresponding to (*iia*) and (*iib*), i.e., it is a parallelogram near the origin. This is in some sense the worst region, particularly if there are many constraints in the integer program. For $[b_1, b_2]$ in this region, (1) we may have no integer solution, and (2) we have troubles of the same nature as noted for (*iia*) and (*iib*).

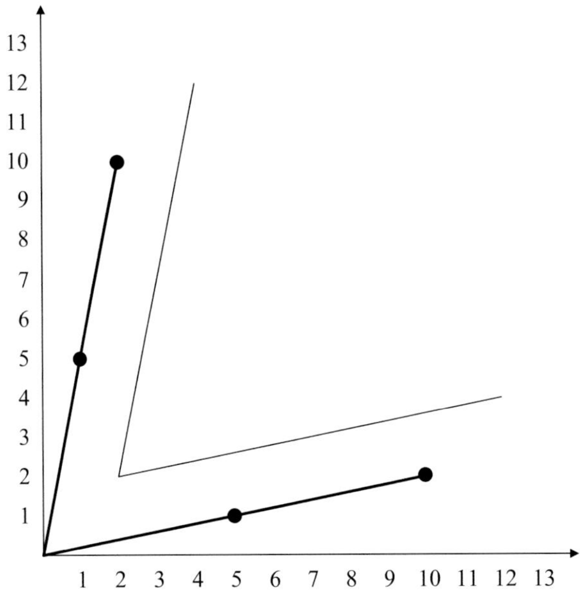

Fig. 10.3 The outer cone and inner cone of the integer program in (10.3) corresponding to basis vectors $\begin{bmatrix} 5 & 1 \\ 1 & 5 \end{bmatrix}$

Since we have no control over the integer program, we may have to find a basis—not the optimal basis of the associated linear program—and try our luck again and again. However, there is some good news. If we randomly select $[b_1, b_2]$ to be our RHS of the integer program, then there are more points in region (i) than all integer points in the other three regions (iia), (iib), and (iii).

If we have an integer program with two constraints, and most vectors have components with small values such as

$$\begin{aligned} \max \quad & x_0 = c_1x_1 + c_2x_2 + \cdots + c_nx_n \\ \text{subject to} \quad & \begin{bmatrix} 5 \\ 1 \end{bmatrix} x_1 + \begin{bmatrix} 1 \\ 5 \end{bmatrix} x_2 + \begin{bmatrix} 1 \\ 1 \end{bmatrix} x_3 + \begin{bmatrix} 2 \\ -1 \end{bmatrix} x_4 + \cdots = \begin{bmatrix} b_1 \\ b_2 \end{bmatrix} \\ & x_j \geq 0 \quad \text{integers} \end{aligned} \tag{10.3}$$

then presumably we do not even have to solve the congruence relations of the associated linear program of (10.3) or worry about its basic variables being forced to be negative. This is because most $[b_1, b_2]$ in (10.3) can be expressed as positive integer combinations of the given vectors in (10.3). In Figure 10.3, we assume that $\begin{bmatrix} 5 & 1 \\ 1 & 5 \end{bmatrix}$ is the optimum basis of the associated linear program of (10.3). Note that the two strips of (10.3) are narrow, the region (iii) is small, and most of the integer points are in region (i).

Linear and Integer Programming in Practice

11

Ultimately, we learn about linear programming and integer programming because we wish to solve real-world problems using these techniques. In this chapter, we first discuss how problems can be formulated as linear and integer programs. We then give example formulation techniques that go well beyond the conversions used to obtain equivalent formulations in Section 4.1. Last, we note several formats in which linear and integer programs are commonly expressed so that modern solver codes can be applied to obtain solutions. Many more examples and exercises are given at this book's website, http://lipme.org.

11.1 Formulating a Real-World Problem

When formulating a problem as a linear program or integer program, the first step is to establish clear and complete notation. Then, three questions must be answered:

1. What is the objective function?
2. What are the optimization variables?
3. What are the constraints?

Once you have answered these three questions, you will be well on your way to writing down the desired linear or integer program.

Example 1: Fitting a Line. Given a set of points with coordinates $(x_1, y_1), \ldots, (x_n, y_n)$, formulate a linear program to find the best-fit linear function, i.e., the line for which the maximum distance of any given point above or below the line is minimized (see Figure 11.1).

To write this as a linear program, the first step is to observe that the general form of a linear function is $ax + by + c = 0$. Thus, to find the best-fit line, we seek the set of values a, b, and c such that the maximum deviation of any given point above or below the line is minimized. If we introduce a variable e (for "error") to denote the

T.C. Hu, A.B. Kahng, *Linear and Integer Programming Made Easy*,
DOI 10.1007/978-3-319-24001-5_11

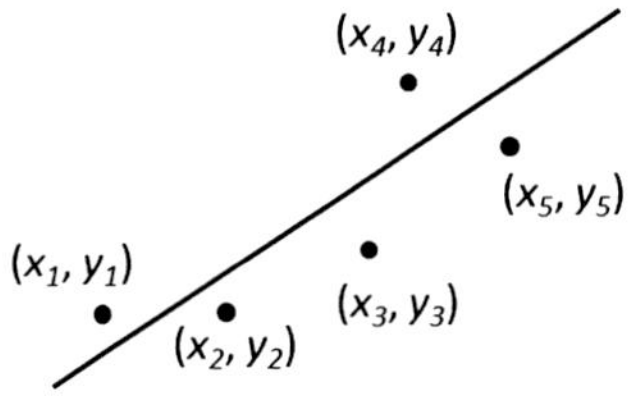

Fig. 11.1 An example to fit a line given five points

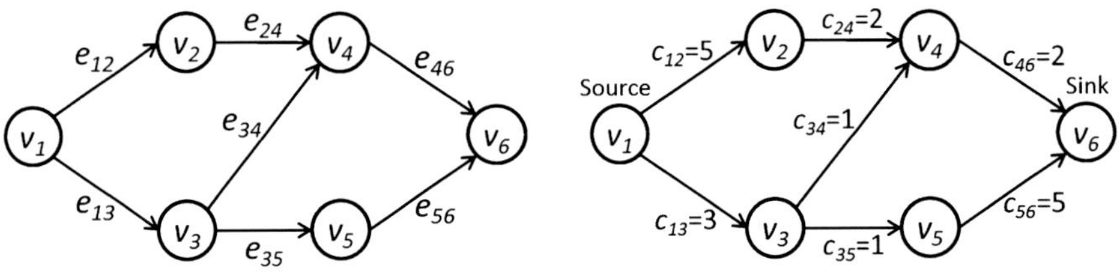

Fig. 11.2 An example of a maximum flow problem

maximum deviation of any given point from the line corresponding to the linear function, then we want to minimize e. We formulate the desired linear program as

$$\begin{aligned} \min \quad & e \\ \text{subject to} \quad & ax_j + by_j + c \leq e \\ & ax_j + by_j + c \geq -e \end{aligned}$$

where there is a pair of inequality constraints for each given point (x_j, y_j).

Example 2: Maximum Source-to-Sink Flow in a Network. A flow network can be considered as a directed graph with n vertices v_j $(j = 1, \ldots, n)$, where v_1 is the identified *source* and v_n is the identified *sink* of the flow, and each e_{ij} is the edge from v_i to v_j. (Imagine a network of pipes, with flows in units of, say, gallons per minute.) Figure 11.2 shows an example with $n = 6$, and seven edges. Each edge e_{ij} of the flow network has a non-negative flow capacity c_{ij}, e.g., $c_{24} = 2$ in the figure. Flows along edges of the network must be non-negative and cannot exceed the corresponding edge capacities. Moreover, flow is conserved at vertices, just as the flow of water is conserved when pipes meet. Thus, at each vertex other than the source and sink, the sum of incoming flows must equal the sum of outgoing flows. Now, for any given flow network, we wish to formulate a linear program that finds the maximum possible flow from source to sink. (In the example of Figure 11.2, the maximum source-sink flow is 3 units, with edges c_{35} and c_{46} being the bottleneck that prevents additional flow.)

To formulate this as a linear program, we first denote the flow on each edge e_{ij} by an optimization variable x_{ij}. Then, we can formulate the linear program as follows:

$$\begin{aligned}
&\max && \sum_j x_{1j} && \text{(amount of flow leaving the source)}\\
&\text{subject to} && x_{ij} \leq c_{ij} && \text{(capacity constraints)}\\
& && \sum_i x_{ij} = \sum_i x_{ji} \quad (j = 1, \ldots, n) && \text{(conservation constraints)}\\
& && x_{ij} \geq 0 && \text{(non-negative flow constraints)}
\end{aligned}$$

Example 3: The Traveling Salesperson Problem. Given a sales territory consisting of n cities, along with the (symmetric) distances d_{ij} between each pair of cities i and j ($1 \leq i, j \leq n$), a traveling salesperson must visit each city exactly once and return to her starting city, using the shortest possible total travel distance. We shall formulate an *integer* linear program to solve the traveling salesperson problem. The integer nature of the optimization arises because we must decide whether edges between cities are either used or not used in the solution: the salesperson cannot "partially use" a given edge. Specifically, we define the binary variable x_{ij} to indicate whether the salesman travels directly from the ith city to the jth city: $x_{ij} = 1$ when the tour uses the i-j edge and $x_{ij} = 0$ otherwise. We also define the integer *artificial variable* t_i, $1 \leq i \leq n$; for any well-formed tour, there exist values for the t_i (essentially corresponding to respective positions of the cities within the tour) that satisfy the constraints below. We may then formulate an integer linear program as follows:

$$\begin{aligned}
&\min && \sum_{ij} d_{ij}x_{ij} && \text{(sum of distances of all edges included in the tour)}\\
&\text{subject to} && \sum_i x_{ij} = 1 \quad (j = 1, \ldots, n) && \text{(only one incoming edge per vertex)}\\
& && \sum_i x_{ji} = 1 \quad (j = 1, \ldots, n) && \text{(only one outgoing edge per vertex)}\\
& && t_i - t_j + nx_{ij} \leq n - 1 && \text{(no subtours = cycles of} < n \text{ vertices)}\\
& && x_{ij} \text{ binary} && \text{(each edge is either used or not used in the tour)}\\
& && t_i \text{ integer} &&
\end{aligned}$$

Example 4: The Graph Coloring Problem. Another integer program arises in contexts that have the flavor of scheduling. For instance, suppose that a college must schedule many classes into available classrooms. If two classes overlap in time, they cannot be scheduled into the same classroom. What is the minimum number of classrooms needed by the college for all its classes to be offered? We may represent each class by a vertex in a graph, and draw an edge between two vertices if the corresponding classes overlap in time. Then, in the resulting graph (V, E), we have a *graph coloring problem*: we seek to assign colors to the vertices of the graph, such that no two adjacent vertices have the same color, and as few colors as possible are used. (Each color is a distinct classroom.) Can we formulate this as an integer linear program? Yes!

One possible formulation introduces the binary variable y_k ($k = 1, \ldots, m$) to indicate whether the kth color is used. We further define a binary "indicator variable" x_{jk} to denote whether the kth color is used on vertex v_j. This leads us to the following integer program:

$$\begin{aligned}
\min \quad & \sum_k y_k \\
\text{subject to} \quad & \sum_k x_{jk} = 1 \quad (j = 1, \ldots, n) && \text{(each vertex has only one color)} \\
& x_{jk} \leq y_k && \text{(color has/has not been used)} \\
& x_{ik} + y_{jk} \leq 1 \quad \left(e_{ij} \in E, \quad k = 1, \ldots, m\right) && \\
& && \text{(adjacent vertices cannot have the same color)} \\
& x_{jk}, y_k \ \text{binary}
\end{aligned}$$

Some readers will realize that this "schedule classes into classrooms" illustration of graph coloring is actually the *interval partitioning* problem, which is greedily solvable in polynomial time. This being said, the general graph colorability problem is intractable and hence germane to the use of integer programming.

Example 5: A Routing ("Multicommodity Flow") Problem. Our final example in this section comes from the design of complex integrated circuits, where electrical connections, called "signal nets", must be routed without touching each other. A *routing graph* in chip design can be viewed as a three-dimensional mesh, as shown in Figure 11.3. Each edge in the routing graph corresponds to an available connection in the {north, west, south, east, up, down} direction. As shown in the figure, the routing of a given net from its "source" pin to its "sink" pin consists of a path of consecutive edges that connects the corresponding source and sink vertices in the routing graph. Then, given the source vertex s_k and sink vertex t_k for each of m two-pin nets, along with the cost of each edge c_{ij} in the routing graph, we seek to route all nets with minimum total cost. This recalls the network flow of Example 2 above, but is inherently a "multi-commodity" flow: since two signals' wires cannot touch each other (much as a flow of water cannot travel in the same pipe as flows of oil or orange juice), our m nets must be routed on disjoint paths.

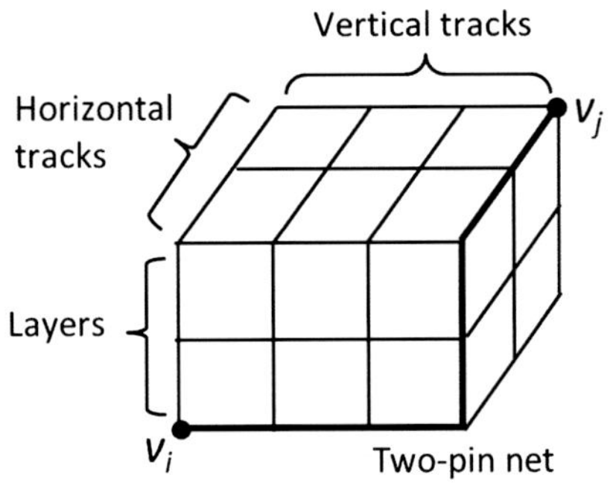

Fig. 11.3 An example of a routing graph

To formulate the routing problem as an integer program, we introduce binary variables x_{ij}^k to indicate whether edge e_{ij} is used in the routing of the kth net. This leads to

$$\begin{aligned}
\min \quad & \sum_k \sum_{ij} c_{ij} x_{ij}^k \\
\text{subject to} \quad & \sum_k \left(x_{ij}^k + x_{ji}^k \right) \leq 1 \qquad \text{(each edge is used by at most one net)} \\
& \sum_i x_{ji}^k - \sum_i x_{ij}^k = \begin{cases} 1, & \text{if } \ v_j = s_k \\ -1, & \text{if } \ v_j = t_k \\ 0, & \text{otherwise} \end{cases} \qquad (k = 1, \ldots, m) \\
& \text{(path structure constraints, analogous to flow conservation)} \\
& x_{ij}^k \ \text{binary}
\end{aligned}$$

The reader may observe recurring motifs in the above integer program examples, notably, the introduction of binary "indicator" variables to represent "used by" versus "not used by" or "assigned to" versus "not assigned to", etc. We next present some additional techniques that are useful in formulating linear and integer programs.

11.2 Tips and Tricks for Expressing Linear and Integer Programs

It is not always obvious how to arrive at a linear objective, or linear constraints, when trying to capture a new optimization problem as a linear or integer program. Absolute value terms, fractional objectives, logical AND or OR operations, etc., can be perplexing. The following are some examples of "tips and tricks", adapted from references [38] and [39]. These can help prevent the size of your linear or integer program (i.e., number of variables or constraints) from exploding unnecessarily—and potentially enable you to find a formulation in the first place. While these examples are not comprehensive, they do provide a flavor of what is possible. For additional "tips and tricks", see this book's website!

Linear Programming

Absolute Values. Suppose that you wish to solve

$$\begin{aligned}
\min \quad & \sum_{j \in J} c_j |x_j| \qquad c_j > 0 \\
\text{subject to} \quad & \sum_{j \in J} a_{ij} x_j \gtrless b_i \quad \forall i \in I \\
& x_j \ \text{free (i.e., unrestricted)}
\end{aligned}$$

Here, the absolute value operator is problematic. However, we can avoid the $|x_j|$ expression by creating two variables x_j^+, x_j^-:

$$x_j = x_j^+ - x_j^-$$
$$|x_j| = x_j^+ + x_j^-$$
$$x_j^+, x_j^- \geq 0$$

Then, we can write down a linear program as follows:

$$\begin{aligned}
\min \quad & \sum_{j \in J} c_j \left(x_j^+ + x_j^- \right) & & c_j > 0 \\
\text{subject to} \quad & \sum_{j \in J} a_{ij} \left(x_j^+ - x_j^- \right) \gtrless b_i & & \forall i \in I \\
& x_j^+, x_j^- \geq 0 & & \forall j \in J
\end{aligned}$$

The original and the rewritten linear programs are the same only if one of x_j^+, x_j^- is always zero for each j. This can be proved by contradiction. If both x_j^+ and x_j^- are positive values for a particular j and $\delta = \min\left(x_j^+, x_j^-\right)$, subtracting δ from x_j^+, x_j^- leaves the value x_j unchanged, but reduces $|x_j|$ by 2δ, which contradicts the optimality assumption.

Min or Max Terms in the Objective. Consider the following optimization:

$$\begin{aligned}
\min \quad & \max_{k \in K} \sum_{j \in J} c_{kj} x_j \\
\text{subject to} \quad & \sum_{j \in J} a_{ij} x_j \gtrless b_i & & \forall i \in I \\
& x_j \geq 0 & & \forall j \in J
\end{aligned}$$

Here, the objective that we want to minimize is the maximum over several terms. This can be accomplished by creating an additional variable z which upper-bounds all the terms over which the max is taken. Hence, our objective becomes the minimization of z. The resulting linear program is then:

$$\begin{aligned}
\min \quad & z \\
\text{subject to} \quad & \sum_{j \in J} a_{ij} x_j \gtrless b_i & & \forall i \in I \\
& \sum_{j \in J} c_{kj} x_j \leq z & & \forall k \in K \\
& x_j \geq 0 & & \forall j \in J
\end{aligned}$$

Maximizing the minimum over several terms can be handled similarly.

Fractional Objective. Consider the following optimization:

$$\begin{aligned}
\min \quad & \left(\sum_{j \in J} c_j x_j + \alpha\right) \Big/ \left(\sum_{j \in J} d_j x_j + \beta\right) & \\
\text{subject to} \quad & \sum_{j \in J} a_{ij} x_j \gtrless b_i & \forall i \in I \\
& x_j \geq 0 & \forall j \in J
\end{aligned}$$

This looks daunting, but, with appropriate manipulation of variables, can also be formulated as a linear program. The first step is to introduce a variable t, where $t = 1/\left(\sum_{j \in J} d_j x_j + \beta\right)$.

The optimization can then be rewritten as

$$\begin{aligned}
\min \quad & \sum_{j \in J} c_j x_j t + \alpha t & \\
\text{subject to} \quad & \sum_{j \in J} a_{ij} x_j \gtrless b_i & \forall i \in I \\
& \sum_{j \in J} d_j x_j t + \beta t = 1 & \\
& x_j \geq 0 & \forall j \in J \\
& t > 0 &
\end{aligned}$$

By introducing additional variables y_j, where $y_j = x_j t$, we obtain the linear program

$$\begin{aligned}
\min \quad & \sum_{j \in J} c_j y_j + \alpha t & \\
\text{subject to} \quad & \sum_{j \in J} a_{ij} y_j \gtrless b_i t & \forall i \in I \\
& \sum_{j \in J} d_j y_j + \beta t = 1 & \\
& y_j \geq 0 & \forall j \in J \\
& t > 0 &
\end{aligned}$$

Finally, by allowing $t \geq 0$ instead of $t > 0$, this linear program is equivalent to the original optimization, provided that $t > 0$ at the optimal solution.

Integer Linear Programming

Variables That Are Minimum Values. Consider the situation where we wish to use a variable that is always equal to the minimum of other variables, e.g.,

$$\begin{aligned} & y = \min\{x_i\} && i \in I \\ & L_i \leq x_i \leq U_i && i \in I \end{aligned}$$

We can realize such a variable y by introducing new binary variables d_i, where

$$\begin{aligned} & d_i = 1 && \text{if } x_i \text{ is the minimum value} \\ & d_i = 0 && \text{otherwise} \end{aligned}$$

The constraints can then be rewritten as follows:

$$\begin{aligned} & L_i \leq x_i \leq U_i && i \in I \\ & y \leq x_i && i \in I \\ & y \geq x_i - (U_i - L_{\min})(1 - d_i) && i \in I \\ & \sum_i d_i = 1 && i \in I \end{aligned}$$

The situation where a variable must always be equal to the *maximum* of other variables, that is, $y = \max\{x_i\}$, is handled analogously (the binary variable $d_i = 1$ if x_i is the maximum value, and $d_i = 0$ otherwise).

Variables That Are Absolute Differences. It may be desirable for a variable to be the absolute difference of two other variables, i.e.,

$$\begin{aligned} & y = |x_1 - x_2| \\ & 0 \leq x_i \leq U \end{aligned}$$

Here, y can be determined by introducing binary variables d_1, d_2 with

$$\begin{aligned} & d_1 = 1 && \text{if} \quad x_1 - x_2 \geq 0 \\ & d_2 = 1 && \text{if} \quad x_2 - x_1 > 0 \end{aligned}$$

Then, the constraints can be rewritten as

$$\begin{aligned} & 0 \leq x_i \leq U \\ & 0 \leq y - (x_1 - x_2) \leq 2 \cdot U \cdot d_2 \\ & 0 \leq y - (x_2 - x_1) \leq 2 \cdot U \cdot d_1 \\ & d_1 + d_2 = 1 \end{aligned}$$

Boolean Functions of Variables. It can also be useful to have a variable that is a Boolean function (e.g., logical AND, OR) of other variables. Notice that the minimum of several binary variables is equivalent to the Boolean AND of the variables: the minimum is 1 only when every variable is equal to 1. Further, the maximum of several binary variables is equivalent to the Boolean OR: the maximum is 1 if any of the variables is equal to 1.

The Boolean AND of several binary variables is given by

$$\begin{aligned} d &= \min\{d_i\} && i \in I \\ d_i &\ \text{binary} && i \in I \end{aligned}$$

which can be rewritten as

$$\begin{aligned} d &\leq d_i && i \in I \\ d &\geq \sum_i d_i - (|I| - 1) && i \in I \\ d &\geq 0 \end{aligned}$$

The Boolean OR of binary variables is

$$\begin{aligned} d &= \max\{d_i\} && i \in I \\ d_i &\ \text{binary} && i \in I \end{aligned}$$

which can be rewritten as

$$\begin{aligned} d &\geq d_i && i \in I \\ d &\leq \sum_i d_i && i \in I \\ d &\leq 1 \end{aligned}$$

The Boolean NOT (negation) of a variable,

$$\begin{aligned} d' &= \text{NOT } d \\ d &\ \text{binary}, \end{aligned}$$

is achieved using

$$d' = 1 - d$$

Constraints with Disjunctive Ranges. Another common scenario is when a variable has a disjunctive range, i.e., it must satisfy the following constraints with $U_1 \leq L_2$:

$$L_1 \leq x \leq U_1 \quad \text{or} \quad L_2 \leq x \leq U_2$$

Introducing a new binary indicator variable

$$y = \begin{cases} 0, & L_1 \leq x < U_1 \\ 1, & L_2 \leq x \leq U_2 \end{cases}$$

allows the above constraints to be rewritten as

$$\begin{aligned} x &\leq U_1 + (U_2 - U_1)y \\ x &\geq L_1 + (L_2 - L_1)y \end{aligned}$$

Fixed-Cost Form of the Objective. Real-world costs often have both fixed and variable components, yielding cost minimization formulations such as

$$\begin{aligned} \text{min} \quad & C(x) \\ \text{subject to} \quad & a_i x + \sum_{j \in J} a_{ij} w_j \gtreqless b_i \quad \forall i \in I \\ & x \geq 0 \\ & w_j \geq 0 \quad \forall j \in J \\ \text{where} \quad & C(x) = \begin{cases} 0, & x = 0 \\ k + cx, & x > 0 \end{cases} \end{aligned}$$

Introducing a new binary indicator variable

$$y = \begin{cases} 0, & x = 0 \\ 1, & x > 0 \end{cases}$$

allows the above constraints to be rewritten as (here, u is a large constant)

$$\begin{aligned} \text{min} \quad & ky + cx \\ \text{subject to} \quad & a_i x + \sum_{j \in J} a_{ij} w_j \gtreqless b_i && \forall i \in I \\ & x \geq 0 \quad \text{integer} \\ & w_j \geq 0 && \forall j \in J \\ & x \leq uy \\ & y \quad \text{binary} \end{aligned}$$

Either-Or and Conditional Constraints. Consider the "either-or" construct

$$\begin{aligned} \min \quad & \sum_{j \in J} c_j x_j \\ \text{subject to} \quad & a_{1j} x_j \leq b_1 && (1) \\ & a_{2j} x_j \leq b_2 && (2) \\ & x_j \geq 0 \quad \forall j \in J \\ \text{where} \quad & \text{one of (1) or (2) } \textit{must hold} \end{aligned}$$

This can be realized by introducing a new binary indicator variable y whose meaning is

$$y = \begin{cases} 0, & (1) \text{ holds} \\ 1, & (2) \text{ holds} \end{cases}$$

The above constraints can then be rewritten as follows, where M_1 and M_2 are large constants:

$$\begin{aligned} \min \quad & \sum_{j \in J} c_j x_j \\ \text{subject to} \quad & a_{1j} x_j \leq b_1 + M_1 y \\ & a_{2j} x_j \leq b_2 + M_2 (1 - y) \\ & x_j \geq 0 \quad \forall j \in J \\ & y \qquad \text{binary} \end{aligned}$$

Closely related is the *conditional constraint*:

$$\begin{aligned} &\text{if} \quad (A) \sum_{j \in J} a_{1j} x_j \leq b_1 \text{ is satisfied} \\ &\text{then} \quad (B) \sum_{j \in J} a_{2j} x_j \leq b_2 \text{ must also be satisfied} \end{aligned}$$

From Boolean algebra, the conditional "A implies B" is logically equivalent to "$(A$ and $\neg B)$ is false", which in turn is equivalent to "$(\neg A$ or $B)$ is true." This is exactly the either-or constraint. With this realization, the conditional constraint becomes

$$\sum_{j \in J} a_{1j} x_j > b_1$$

$$\sum_{j \in J} a_{2j} x_j \leq b_2$$

To convert the strict inequality, the following set of either-or constraints is used, where ϵ is a small constant:

$$\sum_{j \in J} a_{1j} x_j \geq b_1 + \epsilon$$
$$\sum_{j \in J} a_{2j} x_j \leq b_2$$

Product of Variables. A final important case is when we must handle the product of a binary variable d and a continuous, upper-bounded variable $0 \leq x \leq u$. In this case,

$$y = dx$$

can be rewritten as

$$\begin{aligned} & y \leq ud \\ & y \leq x \\ & y \geq x + u(1 - d) \\ & y \geq 0 \end{aligned}$$

11.3 Using Modern Linear and Integer Programming Solvers

We close this chapter by pointing out how one actually obtains solutions to linear and integer programs in practice. Of course, it is not possible to solve real-world mathematical programs by hand, nor is it reasonable to implement the Simplex or other solution methods from scratch. Today, there are many mature, scalable software packages available either for free or for nominal license costs; the references at the end of this book point to examples such as CPLEX from IBM. To input your linear or integer program to a solver, it is necessary to translate the program into one of several standard formats that go by names such as AMPL, GAMS, CPLEX, MPS, etc. Let us see a few examples.

Recall the linear program example (3.7) in Chapter 3:

$$\begin{aligned} \max \quad & v = x_1 + 2x_2 + 3x_3 \\ \text{subject to} \quad & 12x_1 + 12x_2 + \ 6x_3 \leq 30 \\ & \ 4x_1 + 10x_2 + 18x_3 \leq 15 \\ & x_j \geq 0 \end{aligned}$$

We now show representations of this linear program in several different mathematical programming file formats. The website of this book gives additional examples.

Format 1: **AMPL** ("A Mathematical Programming Language"). AMPL frames optimization problems so that they may be fed via AMPL's interface to any of several compatible "back-end" solver codes. To do this, one generally puts the model in a *filename.mod* file and data in a *filename.dat* file. Our example is very simple, so the following *example.mod* file suffices:

```
var x1 >= 0;
var x2 >= 0;
var x3 >= 0;
maximize TOTAL: x1 + 2*x2 + 3*x3;
subject to LIM1: 12*x1 + 12*x2 + 6*x3 <= 30;
subject to LIM2: 4*x1 + 10*x2 + 18*x3 <= 15;
```

Format 2: **GAMS** ("General Algebraic Modeling System"). The same example can be encoded in a file *example.gms* as follows:

```
* Example file example.gms
Free variable TOTAL "TOTAL";
Positive variable x1 "var1";
Positive variable x2 "var2",
Positive variable x3 "var3";

Equations
      obj "max TOTAL"
      lim1 "lim1"
      lim2 "lim2";

      obj .. x1 + 2*x2 + 3*x3 =e= TOTAL;
      lim1 .. 12*x1 + 12*x2 + 6*x3 =l= 30;
      lim2 .. 4*x1 + 10*x2 + 18*x3 =l= 15;

Model example /all/;
Solve example using lp maximizing TOTAL;
```

Format 3: **CPLEX** (a name carried over from a C programming language implementation of the Simplex algorithm). In the CPLEX LP file format, our example becomes

```
Maximize
      TOTAL: x1 + 2 x2 + 3 x3
Subject to
      LIM1: 12 x1 + 12 x2 + 6 x3 <= 30
      LIM2: 4 x1 + 10 x2 + 18 x3 <=15
Bounds
      x1 >= 0
      x2 >= 0
      x3 >= 0
End
```

It can be somewhat tedious to write and debug the small codes or scripts that translate your linear or integer program into one of these input formats. However, such codes and scripts can tremendously improve productivity and the automation of experimental studies—and their elements are highly reusable from one project to another. Further, and finally, you will have a great feeling of satisfaction and accomplishment after having successfully framed your problem as a linear or integer program, correctly feeding the problem to a solver, and obtaining the desired solution!

Appendix: The Branch and Bound Method of Integer Programming

The integer algorithms described in previous chapters are classified as cutting plane type, since they all generate additional constraints or cutting planes. In this section, we shall discuss an entirely different approach, which can be called the *tree search* method. The tree search type of algorithm includes the branch and bound method, the additive algorithm, the direct search algorithm, and many others.

The common features of the tree search type of algorithm are (1) they are easy to understand, (2) they are easy to program on a computer, and (3) the upper bound on the number of steps needed in the algorithm is of the order $O(k^n)$, where n is the number of variables. Features (1) and (2) are two advantages of the tree search type of algorithm. Feature (3) is a disadvantage since it implies exponential growth of the amount of computation as the problem becomes larger.

In the previous chapters, we have concentrated on the cutting plane type algorithms because they give better insight into the problem. For small problems, the tree search type needs less computing time than the cutting plane type, but the growth of the computing time is more rapid in the tree search type.

Consider an integer program

$$\begin{aligned} \min \quad & z = \boldsymbol{c}\boldsymbol{y} \\ \text{subject to} \quad & \boldsymbol{A}\boldsymbol{y} \geq \boldsymbol{b} \\ & \boldsymbol{y} \geq \boldsymbol{0} \quad \text{integers} \end{aligned} \tag{A.1}$$

If each component of $\boldsymbol{y}$ is bounded from above by an integer M, then there are $(M+1)^n$ possible solutions $\boldsymbol{y}$. We could test each of these solutions for feasibility and select the feasible solution with the minimum value of the objective function as the optimum solution. Since the number $(M+1)^n$ is usually very large, the tree search algorithm tries to avoid inspection of solutions that are dominated by solutions already inspected.

In this section, we shall discuss the branch and bound method. We first solve (A.1) as a linear program. If all variables $y_j \geq 0$ and all are integers, then $\boldsymbol{y}$ is clearly the optimum solution to the integer program. If a particular component $y_k = [y_k] + f_k$, where $0 < f_k < 1$, then we solve two linear programs, one with the additional constraint $y_k = [y_k]$ and one with the additional constraint

T.C. Hu, A.B. Kahng, *Linear and Integer Programming Made Easy*,
DOI 10.1007/978-3-319-24001-5

$y_k = \lfloor y_k \rfloor + 1$. If one of the two linear programs, say the one with $y_k = \lfloor y_k \rfloor$, still does not give integer solutions, i.e., $y_l = \lfloor y_l \rfloor + f_l$, then two more linear programs are solved, one with $y_k = \lfloor y_k \rfloor$, $y_l = \lfloor y_l \rfloor$ and one with $y_k = \lfloor y_k \rfloor$, $y_l = \lfloor y_l \rfloor + 1$ as the additional constraints.

All solutions obtained in this way can be partially ordered as a tree with the root of the tree representing the linear program solution obtained without any additional integer constraints. When a solution $\mathbf{y}^0$ does not satisfy the integer constraints, it branches into two other solutions $\mathbf{y}^1$ and $\mathbf{y}^2$. The solution $\mathbf{y}^0$ is called the *predecessor* of $\mathbf{y}^1$ and $\mathbf{y}^2$, and $\mathbf{y}^1$ and $\mathbf{y}^2$are called the *successors* of $\mathbf{y}^0$.

If the successors of $\mathbf{y}^1$ and $\mathbf{y}^2$ are all infeasible, then we have to branch from $\mathbf{y}^0$ again with $y_l = \lfloor y_l \rfloor - 1$ and $y_l = \lfloor y_l \rfloor + 2$. A node may have more than two successors. For example, a node with y_l non-integer may have many successors corresponding to $y_l = \lfloor y_l \rfloor$, $y_l = \lfloor y_l \rfloor - 1$, ..., $y_l = \lfloor y_l \rfloor + 1$, $y_l = \lfloor y_l \rfloor + 2$, ..., etc. A node is called a *terminal node* if it has no successors; this definition implies that a terminal node represents a feasible integer solution or an infeasible integer solution. The idea of the branch and bound method lies in the following two facts:

1. Because the predecessor has fewer constraints than the successors and additional constraints cannot improve the value of the objective function, the optimum value of a successor is always larger than or equal to the optimum value of the predecessor.
2. If two integer feasible solutions have the same predecessor, one with $y_l = \lfloor y_l \rfloor$ $+1$ and one with $y_l = \lfloor y_l \rfloor + 2$, then the optimum value of the first solution is less than the optimum value of the second. This is to say, the further away the value of y_l is from the linear programming solution, the worse is the resulting value of the objective function.

During the computation of the branch and bound method, we keep the optimum value z^* of the best integer feasible solution obtained so far. If a node with a non-integer solution has an optimum value greater than z^*, then all the successors of that node must have optimum values greater than z^*. Hence, there is no sense in branching from that node. The advantage of the branch and bound method is that it can be used for mixed integer problems.

Epilogue

In this epilogue, we hope to give a global view of the book that can serve as a guide to go through the subjects of linear and integer programming again.

Consider a linear program with a single constraint. If we want to maximize the objective function of the linear program, the optimum solution uses variable x_k associated with the item with the largest ratio of its value to its weight.

When a linear program has two constraints, we can use the same intuitive idea, except that the denominator of the ratio is replaced by a 2×2 square matrix. Dividing by a 2×2 square matrix is the same as multiplying by the inverse of the square matrix, and this generalization eventually leads us to the classical Simplex Method for solving m constraints and n columns.

There are several notations used in the first three chapters: min z, max z, or max w and max v. And when we start using the Simplex tableau, we always use max x_0 and min z, say $x_0 + 2x_3 - 2x_4 - x_5 = 0$.

The Simplex Method is shown in Tableaus 4.1, 4.2, and 4.3. A typical Simplex tableau is shown in Tableau 4.1, which shows $x_0 = 0, x_1 = 4$, and $x_2 = 2$. After an iteration, we have $x_0 = 8$, $x_2 = 6$, and $x_4 = 4$ in Tableau 4.2. When all modified coefficients in the top row are non-negative as shown in Tableau 4.3, the solution is optimum.

If we use brute force to solve a linear program, we would need to search over all choices of $(m + 1)$ columns from $(n + 1)$ columns, and for each choice solve the simultaneous equations. In the Simplex Method, we choose the column with the most negative coefficient to enter the basis, and then do a pivot step. The pivot element a_{rs} in the rth row and s^{th} column is decided by a feasibility test.

If the modified coefficient is already non-negative, we do not need to find all entries in that column. Also, we do not need to carry all entries, for iterations after iterations.

The revised Simplex Method is introduced in Chapter 6 with Tableaus 6.1, 6.2, 6.3, 6.4, 6.5, and 6.6. The entries that are not calculated are replaced by question marks. To show the savings of the revised Simplex Method, in the revised savings

T.C. Hu, A.B. Kahng, *Linear and Integer Programming Made Easy*,
DOI 10.1007/978-3-319-24001-5

of the revised Simplex Method, we basically carry entries in a tableau of $(m+1) \times (m+1)$ entries. Roughly speaking, when a linear program has m rows and n columns, it pays to use the revised Simplex Method when $3 < m < \frac{n}{3}$.

When n is too large to write down, we are forced to use the column generating technique. In the column generating technique, the problem of selecting the best column to replace one of the existing columns in the basis becomes an optimization problem.

The simplest integer program is the knapsack problem discussed in Chapter 8. A simple example is

$$\begin{aligned} \text{min} \quad & z = 10x_1 + 8x_2 \\ \text{subject to} \quad & 10x_1 + 7x_2 \geq 101 \\ & x_1, x_2 \geq 0 \end{aligned}$$

We could try

$$\begin{aligned} x_1 &= 10 \quad \text{and} \quad x_2 = 1 \quad \text{with} \quad z = 108 \\ x_1 &= 9 \quad \text{and} \quad x_2 = 2 \quad \text{with} \quad z = 106 \\ & \cdots \\ x_1 &= 8 \quad \text{and} \quad x_2 = 3 \quad \text{with} \quad z = 104 \end{aligned}$$

A key observation is that an integer program with two constraints is easier to understand, as shown in Chapter 9. When we drop the integer restriction on variables, the associated linear program has many column vectors with $x_1 = 2.75$, $x_2 = 2.75$, $x_3 = 0$, $x_4 = 0$, and $z = 5.5$, which is less than the value $z = 6.0$ obtained by $x_1 = 2$, $x_2 = 2$, $x_3 = 1$, and $x_4 = 1$. The idea is to use the non-basic variables x_3 and x_4 as little as possible.

Assume we have an integer program

$$\begin{aligned} \text{min} \quad & z = x_1 + x_2 + x_3 + x_4 + x_5 + x_6 \\ \text{subject to} \quad & \begin{bmatrix} 3 \\ 1 \end{bmatrix} x_1 + \begin{bmatrix} 1 \\ 3 \end{bmatrix} x_2 + \begin{bmatrix} 2 \\ 1 \end{bmatrix} x_3 + \begin{bmatrix} 1 \\ 2 \end{bmatrix} x_4 + \begin{bmatrix} 1 \\ 0 \end{bmatrix} x_5 + \begin{bmatrix} 0 \\ 1 \end{bmatrix} x_6 = \begin{bmatrix} 11 \\ 11 \end{bmatrix} \\ & x_j \geq 0 \end{aligned}$$

Here, there are several associated linear programs:

$$\text{with} \quad \begin{cases} x_1 = \dfrac{11}{4} \\ x_2 = \dfrac{11}{4} \end{cases} \quad \text{and} \quad z = \frac{11}{4}$$

$$\text{and} \quad \begin{cases} x_3 = \dfrac{11}{3} \\ x_4 = \dfrac{11}{3} \end{cases} \quad \text{and} \quad z = \frac{22}{3}$$

$$\text{and} \quad \begin{cases} x_5 = 11 \\ x_6 = 11 \end{cases} \quad \text{and} \quad z = 22$$

When we raise the values of non-basic variables, we may force the basic variables to be negative. In general, we could take the optimum solutions of the associated linear program and combine them with non-basic variables that satisfy the congruence relationships. However, it is possible that a better basis of the associated linear program may not give a better value of the objective function.

References

References on Linear Programming

1. Bland RG (1977) New finite pivoting rules for the simplex method. Math Oper Res 2 (2):103–107
2. Dantzig GB, Orchard-Hayes W (1953) Alternate algorithm for revised simplex method using product form of the inverse. RAND Report RM 1268. The RAND Corporation, Santa Monica
3. Dantzig GB (1963) Linear programming and extensions. Princeton University Press, Princeton
4. Dantzig GB, Thapa MN (1997) Linear programming. In: Introduction, vol 1; Theory and extensions, vol 2. Springer series in operations research. Springer, New York
5. Hu TC, Shing M-T (2001) Combinatorial algorithms. Dover, Mineola
6. Hu TC (1969) Integer programming and network flows. Addison-Wesley, Reading
7. Lemke CE (1954) The dual method of solving the linear programming problems. Naval Res Log Q 1(1):36–47
8. Von Neumann J, Morgenstern O (1944) Theory of games and economic behavior. Wiley, New York
9. Wolfe P (1963) A technique for resolving degeneracy in linear programming. J SIAM Appl Math 11(2):205–211

References on Integer Programming

10. Araoz J, Evans L, Gomory RE, Johnson EL (2003) Cyclic group and knapsack facets. Math Program Ser B 96:377–408
11. Chopra S, Jensen DL, Johnson EL (1983) Polyhedra of regular p-nary group problems. Math Program Ser A 43:1–29
12. Gomory RE, Hu TC (1962) An application of generalized linear programming to network flows. J SIAM 10(2):260–283
13. Gomory RE, Hoffman AJ (1962) Finding optimum combinations. Int Sci Technol 26–34
14. Gomory RE, Hoffman AJ (1963) On the convergence of an integer programming process. Naval Res Log Q 10(1):121–123
15. Gomory RE (1963) All-integer integer programming algorithm. In: Muth JF, Thompson GL (eds) Industrial scheduling. Prentice Hall, Englewood Cliffs, pp 193–206
16. Gomory RE (1963) An algorithm for integer solutions to linear programs. In: Graves RL, Wolfe P (eds) Recent advances in mathematical programming. McGraw-Hill, New York, pp 269–302

T.C. Hu, A.B. Kahng, *Linear and Integer Programming Made Easy*,
DOI 10.1007/978-3-319-24001-5

17. Gomory RE (1963) Large and non-convex problems in linear programming. In: Proceedings of the symposium on the interactions between mathematical research and high-speed computing of American Mathematics Society, vol 15, pp 125–139
18. Gomory RE, Hu TC (1964) Synthesis of a communication network. SIAM J 12(2):348–369
19. Gomory RE, Johnson EL (2003) T-space and cutting planes. Math Program Ser B 96:341–375
20. Gomory RE (1965) Mathematical programming. Am Math Mon 72(2):99–110
21. Gomory RE (1965) On the relation between integer and non-integer solutions to linear programs. Proc Natl Acad Sci U S A 53(2):260–265, Also published in Dantzig GB, Veinott AF Jr (eds) (1968) Mathematics of the decision sciences, Part 1. Lectures in applied mathematics, vol 2. American Mathematical Society, pp. 288–294
22. Gomory RE (1969) Some polyhedra related to combinatorial problems. J Linear Algebra Appl 2(4):451–558
23. Gomory RE, Johnson EL (1972) Some continuous functions related to corner polyhedra, part I. Math Program 3(1):23–85
24. Gomory RE, Johnson EL (1972) Some continuous functions related to corner polyhedra, part II. Math Program 3(3):359–389
25. Gomory RE, Baumol WJ (2001) Global trade and conflicting national interest. MIT Press, Cambridge
26. Gomory RE, Johnson EL, Evans L (2003) Corner polyhedra and their connection with cutting planes. Math Program 96(2):321–339
27. Hu TC (1969) Integer programming and network flows. Addison-Wesley, Reading, Chapter 17 by R. D. Young
28. Hu TC, Landa L, Shing M-T (2009) The unbounded knapsack problem. In: Cook WJ, Lovász L, Vygen J (eds) Research trends in combinatorial optimization. Springer, Berlin, pp 201–217
29. Johnson EL (1980) Integer programming: facets, subadditivity and duality for group and semi-group problems. CBMS-NSF conferences series in applied mathematics 32
30. Schrijver A (1986) Theory of linear and integer programming, Wiley interscience series in discrete mathematics and optimization. Wiley, New York
31. Schrijver A (2002) Combinatorial optimization. Springer, Berlin, Vol A Chapters 1–38, pp 1–648; Vol B Chapters 39–69, pp 649–1218; Vol C Chapters 70–83, pp 1219–1882
32. Young RD (1965) A primal (all integer) integer programming algorithms. J Res Natl Bur Stand B 69(3):213–250, See also Chapter 17, pp 287–310 in the book by T. C. Hu

References on Dynamic Programming

33. Bellman RE (1956) Dynamic programming. R-295 RAND Corporation. The RAND Corporation, Santa Monica
34. Bellman RE, Dreyfus SE (1962) Applied dynamic programming. R-352-PR RAND Corporation. The RAND Corporation, Santa Monica
35. Dreyfus SE, Law AM (1977) The art and theory of dynamic programming, vol 130, Mathematics in science and engineering. Academic, New York

References on Practical Formulation of Linear and Integer Programs

36. AIMMS modeling guide—integer programming tricks (Chapters 6, 7). http://download.aimms.com/aimms/download/manuals/AIMMS3_OM.pdf
37. Fair Isaac Corporation (2009) MIP formulations and linearizations: quick reference. http://www.fico.com/en/node/8140?file=5125
38. Fourer R, Gay DM, Kernighan BW (1990) A modeling language for mathematical programming. Manage Sci 36:519–554

39. ILOG CPLEX Optimizer. http://www-01.ibm.com/software/commerce/optimization/cplex-optimizer/ (see also https://en.wikipedia.org/wiki/CPLEX)
40. NEOS Optimization Server. http://www.neos-server.org/neos/solvers/
41. GAMS: A User's Guide. http://www.gams.com/help/topic/gams.doc/userguides/GAMS UsersGuide.pdf

References on NP-Completeness

42. Cook SA (1973) A hierarchy for nondeterministic time complexity. J Comput Syst Sci 7(4):343–353
43. Garey MR, Johnson DS (1979) Computers and intractability: a guide to the theory of NP-completeness. Freeman, San Francisco
44. Karp RM (1972) Reducibility and combinatorial problems. In: Miller RE, Thatcher JW (eds) Complexity of computer computation. Plenum, New York, pp 85–103

Index

T.C. Hu, A.B. Kahng, *Linear and Integer Programming Made Easy*,
DOI 10.1007/978-3-319-24001-5